An Introduction to
MICROSCOPY

An Introduction to
MICROSCOPY

Suzanne Bell
Keith Morris

CRC Press
Taylor & Francis Group
Boca Raton   London   New York

CRC Press is an imprint of the
Taylor & Francis Group, an **informa** business

CRC Press
Taylor & Francis Group
6000 Broken Sound Parkway NW, Suite 300
Boca Raton, FL 33487-2742

© 2010 by Taylor and Francis Group, LLC
CRC Press is an imprint of Taylor & Francis Group, an Informa business

Printed in the United States of America on acid-free paper
10 9 8 7 6 5 4 3 2 1

International Standard Book Number: 978-1-4200-8450-4 (Hardback)

Library of Congress Cataloging-in-Publication Data

Bell, Suzanne.
    An introduction to microscopy / Suzanne Bell, Keith Morris.
        p. cm.
    Includes bibliographical references and index.
    ISBN 978-1-4200-8450-4 (hardcover : alk. paper)
    1. Microscopy. I. Morris, Keith, 1965- II. Title.

QH205.2.B45 2010
502.8'2--dc22   2009031928

Visit the Taylor & Francis Web site at
http://www.taylorandfrancis.com

and the CRC Press Web site at
http://www.crcpress.com

Contents

Preface

Microscopy as an academic subject is nearly a lost art, yet it has served as a fundamental scientific technique for centuries. Indeed, for hundreds of years, it was arguably the *only* scientific method. It remains an invaluable tool in biology and healthcare and has been integrated increasingly into modern chemical instrumentation. For example, microscope-based spectrometers operating in the ultraviolet, infrared, and visible ranges are readily available, made feasible by advances in laser and detector technology; all such devices operate as microscopes. To understand these instruments, it is critical to master the foundational principles of microscopy. The goal of this textbook is to assist you in that process.

This book is designed for use in an undergraduate or graduate course in microscopy and as a stand-alone reference that professionals can use for self-study. All nine chapters include questions and most include simple exercises related to the material covered. Numerous figures and photographs supplement the text and explain the procedures and principles introduced. Although we used microscopes in our laboratories to produce these images, we present the material in a general way so that it can be applied to any type of microscope to which you have access.

We assume that you have taken mathematics courses through algebra and have a basic knowledge of physics (essential for understanding optics) and chemistry. This book is intentionally concise and we do not attempt to cover all aspects of all types of microscopy such as polarizing light and fluorescence. Rather, our intent is to provide you with the basic knowledge necessary to explore and understand these more advanced techniques. The appendices contain a detailed bibliography and selection of references to supplement your study and, where appropriate, we supply references within chapter text. Think of this book as a gateway, a text to accompany *Microscopy 101*. Enjoy!

Authors

Dr. Suzanne Bell earned a BS with a dual major in chemistry and police science (criminal justice) at Northern Arizona University, followed by an MS in forensic science from the University of New Haven. After an internship with the New Mexico State Police, she began work there in 1983. Her main duties involved forensic chemistry, but she also processed numerous crime scenes and worked many arson and trace evidence cases. She then spent 10 years in the Organic Analysis group at the Los Alamos National Laboratories and supervised the group for 2 years before returning to school. She obtained a PhD in chemistry from New Mexico State University in 1991 and completed a post-doctoral fellowship in 1994. She then accepted a faculty position at Eastern Washington University where she worked closely with the Washington State Patrol to launch a forensic program leading to a BS in chemistry with a forensic science emphasis.

Dr. Bell joined the faculty of West Virginia University in 2003 and works in both the forensic chemistry and forensic identification aspects of the program at the undergraduate and graduate levels. She developed a forensic chemistry lecture and laboratory course (400 level) and a textbook entitled *Forensic Chemistry* (Pearson/Prentice Hall, 2005). Other books include *Crime and Circumstance* (Greenwood/Pragear, 2008), and for Facts-on-File, *The Encyclopedia of Forensic Science* (First and Second Editions), and *The Dictionary of Forensic Science*. She served as series editor of *The Essentials of Forensic Science* and authored two volumes in the series, *Drugs and Poisons* and *Fakes and Forgeries*.

Dr. Bell is a diplomate of the American Criminalistics Association and a member of the American Academy of Forensic Science.

Dr. Keith Morris holds a BSc (hons) and PhD from the University of Port Elizabeth (South Africa), and a master's degree in business leadership from the University of South Africa. He was attached to the Forensic Science Laboratory of the South African Police Service for 14 years, during which time he specialized in casework involving trace evidence, precious metal thefts, and bombing investigations. He was also responsible for training detectives and crime scene examiners to preserve, collect, and analyze forensic evidence. His work on major cases included the bombing of the Planet Hollywood restaurant in Cape Town in 1998. For 6 years of his tenure with the Police Service, he served as director of the Forensic Science Laboratory System, and was decorated with the Police Star for Outstanding Service.

Dr. Morris is currently the director of the Forensic and Investigative Science Program at the Eberly College of Arts and Sciences at West Virginia University. The program is nationally accredited through FEPAC. He was instrumental in establishing a master's level forensic and investigative science program at the university and teaches extensively in that program.

Dr. Morris is a member of the General Forensics Technical Working Group of the National Institute of Justice, the AFIS Expert Group, and the Human Factors in Latent Print Analysis Group of NIST/NIJ. He is also a member of the American Academy of Forensic Science and the International Association for Identification. In 2008, Dr. Morris was named the Ming Hsieh Distinguished Professor in the Forensic and Investigative Science Program at West Virginia University.

Introduction

Microscopic characterization is a powerful analytical tool available across many scientific disciplines. By observing what at first seems a simple process of the interaction of visible light with matter, a skilled microscopist can identify crystal structures and unknown substances, characterize minerals, determine disease processes and damage, and perform advanced spectroscopic measurements using visible, infrared, Raman, ultraviolet fluorescence, and scanning electron microscopy instruments. The microscope keeps evolving, improving, and expanding its utility. Understanding the fundamentals of microscopy constitutes a significant asset in the workplace, as many scientists work with microscopes even if they have not had significant academic training in the subject.

At its core, microscopy is a tool for the exploration of the interaction of electromagnetic energy (here, primarily visible light) with matter. This definition applies equally well across the electromagnetic spectrum. One way to think of microscopy is as visible light spectroscopy in which your eye or a camera functions as the detector. If you understand the fundamentals of how a microscope works, how light is moved and manipulated throughout the optic train, and how it interacts with a sample to reveal specific information, you will have taken a significant step toward understanding spectroscopy. For example, the relationship between slit widths and intensity is easy to understand because you can literally see this relationship through a microscope. If you grasp the concepts behind polarizing light microscopy, you will understand beam splitters and how Fourier transform-based interferometers work. These devices are at the heart of many modern analytical instruments.

Our treatment begins with some fundamentals of light and moves quickly into the basics of optics and optical trains utilized in microscopes. Whole chapters are devoted to Kohler Illumination and photography through a microscope—two related topics that cover skills central to the modern practice of microscopy. We discuss sample handling and preparation from a general perspective and focus on the sample as dictating the preparation method. Later chapters build on this information, delving into crystallography and polarizing light microscopy, and finally chemical microscopy. You may use any chapter as a stand-alone reference, of course, but the flow of this book is organized intentionally and we recommend going through it from start to finish. We also provide detailed appendices that include answers to chapter questions, abbreviations, a detailed glossary, and a bibliography that lists resources we used in writing this book as well as general references currently used in microscopy. Based on the ephemeral nature of Internet sites, we purposely limited references to those that have demonstrated staying power and appear likely to persist for the useful lifetime of this book.

We begin then with Chapter 1 that covers the fundamentals of light.

1 Light and Matter

Microscopes are devices that move light from a source, through a sample where interaction takes place, and then to the eye or a camera. In later chapters, we will explore the design and construction of microscopes and discuss the basics of optics. To understand these topics, you must understand the fundamentals of light and the terms associated with it.

In a sense, microscopes are colorimeters that use the eye as a detector. Accordingly, much of our discussion can be extended to spectroscopy in general. Later, you will learn about how polarized light is split by certain types of materials into two beams that recombine at the detector. This same concept can be used to understand how beam splitters in spectrometers such as Fourier transform infrared spectrometers (FTIRs) operate. Concepts such as aperture and intensity apply as much to microscopy as to spectroscopy. The advantage of using microscopy to explore light is that we can look through a microscope and literally see how light interacts with matter. The microscopist functions as a detector and is able, literally and figuratively, to visualize fundamental properties and interactions.

1.1 DUAL NATURE OF LIGHT

What we call *light* is the visible portion of the larger **electromagnetic spectrum** (**EMS**) of energy shown in Figure 1.1. The term *electromagnetic radiation* (EMR) is sometimes used. *Radiation* can be a confusing designation, so we will avoid it here. *Light* will be used interchangeably with *energy* in this book, with the understanding that our emphasis will be on the visible range of the spectrum. Two models are used to describe **electromagnetic energy**: the wave model and the particle model. These are not mutually exclusive; light behaves simultaneously as a wave and a particle. The application dictates which terminology is most appropriate and you must be familiar with both.

In the particle model (Figure 1.2, left), the light source emits discrete energy packets called **photons**, whose energies are calculated as shown. In the wave model (Figure 1.2, right), light consists of a wave propagating through space. A useful analogy is a bowling ball dropped into the middle of a pond. Waves will propagate outward from the drop point through the water and will reach the shore with measurable wavelengths (distance from crest to crest) and frequency. In this analogy, frequency is defined as the number of waves that impact the shore per second. Returning to light emitted from a source, the speed of propagation through a vacuum is taken as a constant C with the relationship between the wavelength and frequency as shown in the figure. The two equations can be linked through frequency (units of Hz or sec^{-1}). As we will see later, the speed of light changes depending on the environment, resulting in refraction. Without this simple phenomenon, there would be no microscopy.

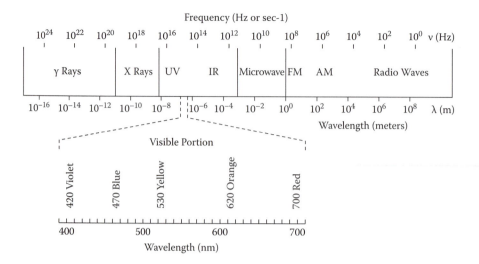

FIGURE 1.1   Spectrum of electromagnetic energy. The visible portion is one small part of this continuum.

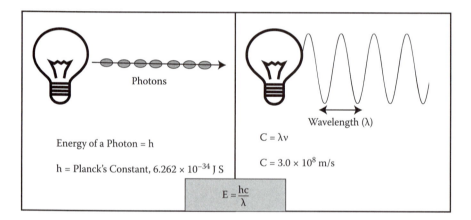

FIGURE 1.2   Comparison of particle and wave models.

The wave shown in Figure 1.2 is simplified. As shown in Figure 1.3, the energy wave propagating through space has two components: electrical and magnetic. These components vibrate perpendicularly to each other. Often the wave is depicted in one dimension instead of two, which is perfectly acceptable in this context. In microscopy, our interest is with the electrical component, not the magnetic one. The direction of propagation is the direction the wave is moving in space; the direction of vibration is the up-and-down wave movement. When we discuss polarized light microscopy in Chapter 7, we will see that waves (beams) of energy can be split into parallel and perpendicular components, but this is not the same concept we are discussing here. Figure 1.3 depicts the dual electrical and magnetic properties of energy.

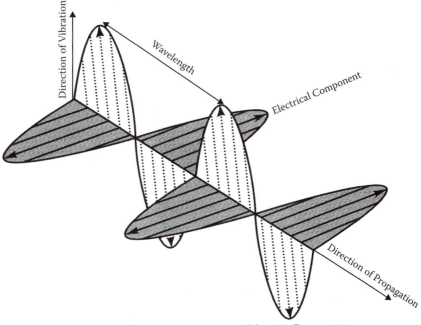

Direction of Vibration

Wavelength

Electrical Component

Direction of Propagation

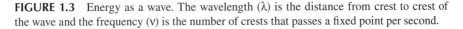

Magnetic Component

FIGURE 1.3 Energy as a wave. The wavelength (λ) is the distance from crest to crest of the wave and the frequency (ν) is the number of crests that passes a fixed point per second.

In the earliest microscopes, the source of light used to illuminate the sample was the sun, a candle, or other indoor light. The light was directed into the instrument using a manually adjustable mirror. Modern microscopes have built-in light sources that may be aligned for optimum illumination. Most microscopes use lamps with tungsten filaments as light sources. Current is applied to the filament, causing it to glow. This light illuminates the samples. Admittedly, this is a superficial description of a complex process, and the rest of this book is geared to filling out this description. An understanding of this process is essential to understanding how light interacts with matter which is, after all, the underlying process exploited in a microscope.

Electromagnetic energy is created by the accelerated motions of electrons that are modeled here as discrete charged particles. Because they are charged, electrons have associated electrical fields. If an electron could be frozen in time and space, its electrical field would be symmetrical and static. Now suppose the electron was suddenly accelerated. The field would not react instantly; rather the field closest to the electron would react immediately while the regions of the field farther away would take longer to react. Consider a simple tungsten wire that might be used in a microscope lamp. The electrons in the wire are not isolated; they are constrained within the orbital structure of the conductive metal. Thermal excitation (resistive heating) causes the atoms in the wire to move, resulting in interactions between metal nuclei and electrons. The result is a complex three-dimensional pattern of electron acceleration and deceleration, displacement, and return. In a tungsten wire, electrons in the valence band will oscillate and emit light. The local environment for any of these electrons changes constantly and thus changes the nature of the oscillation as well. As a result, the wire emits a range of wavelengths and frequencies of light.

For any given oscillating electron, the amplitude of the electromagnetic wave is maximized perpendicularly to the direction of the displacement of the electron. As we will see later, this means that the light emitted by a single particle is polarized (vibrating in a single plane).

Keep in mind that this is a simplistic, two-dimensional view of a complex three-dimensional phenomenon; still, it provides us with a solid basis for understanding what happens when light interacts with matter and why.* At the fundamental level, when an oscillating electric field encounters atoms (positively charged nuclei and negatively charged electron clouds), things (electromagnetic interactions of repulsion and attraction) happen. A microscope can allow us to see the effects of this interaction by observing changes in the light brought on by interaction with matter. Thus, we can link what we see back to fundamental information about the structure of the matter in our sample. Broadly speaking, this is true of all types of spectroscopy.

1.2 VECTOR NOTATION AND WAVEFRONTS

In many types of microscopy and particularly in polarizing light microscopy, it is convenient to use vector notation to depict waves and electromagnetic energy. As shown in Figure 1.4, an observer looking at a wave as it approaches would see the wave vibrating up and down in the plane of vibration and this can be depicted as shown in the figure. We can extend this notation to more than one plane of vibration, as shown in Figure 1.5. At any point, contributions will be made by the wave vibrating in the x plane and the wave vibrating in the y plane. These can be denoted as vectors and combined using vector addition, which takes into account the magnitudes and directions of both vectors.

At the highlighted point in Figure 1.5, line 1 is aligned along one direction and line 2 in a perpendicular direction. Note that for line 1, the amplitude is about two thirds of maximum, and the same is true of line 2. The circle notation illustrates these magnitudes relative to the maximum, which is illustrated as in Figure 1.4. The addition product is shown as a gray line in the lower right of Figure 1.5 and corresponds to the black arrow emanating from the highlighted point. Understanding vector notation is essential for studying polarizing light microscopy (Chapter 7).

When light waves vibrating in a single plane interact, **interference** results. This interference may be constructive or destructive. How these processes are depicted depends upon whether we use the wave or the particle model. For the particle model,

---

* We are discussing only the electric portion of the electromagnetic component here. The magnetic component is always present when a charge moves, but our interest here is with the electronic portion and how it interacts with matter.

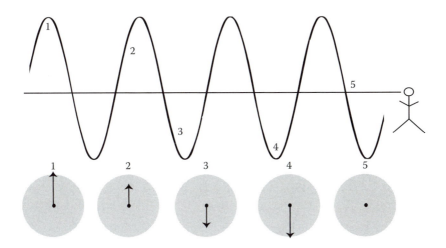

FIGURE 1.4 Vector notation for a wave. An observer is looking down the axis of propagation and sees the movement shown in the circles. At point 1, the wave crest (maximum positive amplitude) has been reached. At point 4, the wave has reached maximum negative amplitude relative to the axis. At point 5, the wave is exactly on the axis and appears as a dot to an observer. The other points demonstrate positions between these extremes.

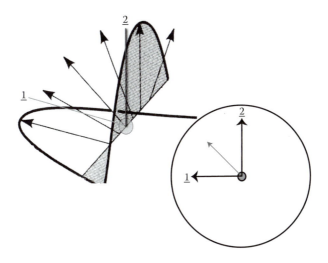

FIGURE 1.5 Vector addition of two waves.

consider again the analogy of dropping a bowling ball into a still pond, but now imagine dropping two bowling balls close to each other at the same instant. The waves emanating from the two drop points will collide and interact to generate a new wave pattern. In some cases, the resulting wave will be larger (reinforced), and in others, the resulting wave will be **attenuated** or even canceled out completely. What occurs depends on how the waves are aligned when they combine.

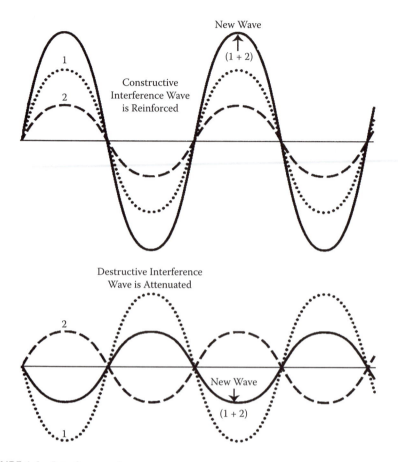

FIGURE 1.6   Interference of waves.

Figure 1.6 shows the results of different interference interactions. In the top frame, the two waves reinforce each other and the resulting wave has higher amplitude than either of the originals. This occurs when the waves are in phase, i.e., vibrating in the same direction at the same time. Conversely, when the waves are out of phase (bottom frame), the resulting wave has lower amplitude than either of the originals. Keep in mind that this is a simplified description; any number of waves can interfere with each other. In the case of a microscope with a **polychromatic** light source, many wavelengths coexist and interact throughout the instrument. In polarizing light microscopy, the colors you see are actually produced by interference interactions and do not represent the color of the sample.

Constructive and destructive interferences are integral to another key concept in microscopy: **diffraction**. When light is directed through a small enough aperture, the wave is disrupted at the edges and curves "backward" toward the edge of the opening, as shown in Figure 1.7. A person viewing the light sees a bright spot in the center of the aperture surrounded by a series of concentric rings of lesser intensity.

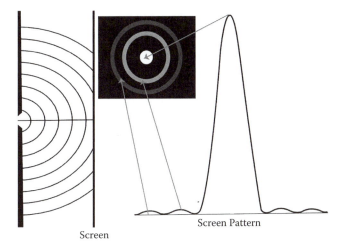

Screen

Screen Pattern

FIGURE 1.7 Single slit experiment. Light is directed through a small slit, creating a diffraction pattern on the screen.

Each ring appears as a result of constructive interference that occurs at whole number multipliers of the wavelength of interest. Again, this is a simplified depiction of a single waveform. A microscope has a number of apertures through which light passes and diffraction inevitably occurs at each aperture. We will discuss this more when we talk about the design, construction, and components of microscopes in Chapters 3 and 4.

1.3   INTERACTIONS OF LIGHT AND MATTER

Light interacts with itself and with matter; both interactions are important to a microscopist. Constructive and destructive interactions and diffraction are important light-with-light interactions; now we will look at how light interacts with matter. First, consider a sample in a light path such as shown in Figure 1.8. Several events can occur when light impinges on the sample, and several may occur at once. What happens depends on the physical and chemical properties of the sample as well as its morphology. In microscopy, this last factor may be particularly important.

In some cases, light is simply reflected back off the sample. Reflection can be **specular**, i.e., the type of reflection seen in a mirror. The angle of reflection is the same as the angle of incidence when compared to a line that is normal to the surface. Diffuse reflectance occurs when a sample surface is not smooth and light is scattered and reflected. To visualize the concept, imagine scratching a mirror's surface with sandpaper and observing the results. Another example is the difference between glossy (reflective) and matte (diffuse) paint finishes. A non-shiny finish is creating by formulating the paint with particulates designed to scatter the impinging light.

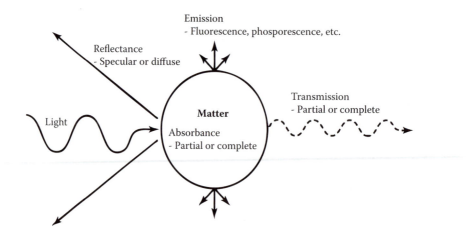

FIGURE 1.8   Possible interactions between light and matter.

If light enters the sample, another set of interactions becomes possible. First, the light may not interact at all and will be transmitted unchanged through the sample. Partial absorbance may occur; in this case the transmitted wave is attenuated by the interaction. This is the basis of many types of spectroscopy in which the patterns of absorbance are used to understand the chemical composition of the sample. In other cases, the absorbed light may be sufficient to cause electron promotion and excitation within the sample, resulting in the emission of energy of the same or different wavelengths. Fluorescence is an example, and fluorescence microscopy has become an important scientific tool in its own right. Notice that when a sample emits energy, it can come from any direction, whereas transmission occurs along the same axis as the beam of impinging light.

The same interactions are possible at a surface, as shown in Figure 1.9. Assume for our purposes that the surface is made of glass, e.g., a lens or a microscope slide. Once again, reflection and absorbance may occur. Any light entering the glass will be subject to **refraction**, a topic we will discuss in detail in the next chapter. Refraction occurs because the speed of light changes as it moves from one medium or environment to another.

Light may also penetrate a small distance into the surface before being reflected back, and in some cases a series of internal reflections occur. This concept serves as the basis of several technologies, including attenuated total reflectance infrared spectroscopy (ATR-IR). Again, both figures depict simple situations with a single wavelength of light. In microscopy, the source lamp emits polychromatic light and the interactions shown here depend on (among other factors) wavelength. This can be an advantage or a disadvantage. For example, in designing lenses, manufacturers must address **chromatic aberration** that arises because different wavelengths of light have different refractive indices. However, this same property allows the design of prisms that disperse light based on wavelength.

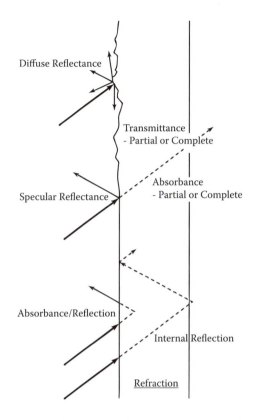

FIGURE 1.9 Possible interactions between light and matter at a surface.

QUESTIONS

1. When light leaves one medium (such as air) and enters another (such as water), the speed of propagation changes. Why? Explain on the molecular level.
2. Based on the concept of diffraction as illustrated in Figure 1.7, what are the implications for designing microscope lenses?

EXERCISES

Exercise 1.1: Diffraction Patterns

Materials needed:

Stiff cardboard (at least 1 mm thick), ~10 pieces, each about 4″ × 4″
Small intense flashlight
Laser pointer (red; other colors if available)
Paper clips (unwound), toothpicks, etc.
Pins and needles

Each time light passes through small apertures in microscopes, diffraction occurs. You can create diffraction patterns using the simple tools above. For the first set of experiments, punch a single hole into the center of a piece of cardboard using different sized implements such as toothpicks, paper clips, sewing needles, and pins. Make sure to work the cardboard out of the light path to make a clear hole. Darken a room or closet and sit near a wall. Hold the light source a few millimeters away from the cardboard and the cardboard a few inches away from the wall. Shine light through the holes and record your observations. This part takes patience. The room should be as dark as possible, complicating such a seemingly simple operation. You should be able to reproduce a pattern similar to Figure 1.7.

For the second round, make a second hole of the same size next to the first in each piece of cardboard. The holes should be as near as possible without overlap. Repeat the lighting and observation steps, then repeat the entire sequence with three holes. Use wavefront diagrams and references to explain the patterns you observed. What differences did you observe between the white light pattern (as in a microscope) and the pattern you saw with a laser pointer that is essentially monochromatic?

2 Fundamentals of Light and Optics

When faced with the challenge of analyzing a sample with a microscope, the obvious question is how to set up a microscope to achieve the required result. Different applications require different set ups. Three fundamental requirements must be addressed. These are **magnification**, **resolution**, and **contrast**. To achieve the best performance from our system (the microscope), we must understand its basic parts and how each of these parts interacts with light.

To apply microscopy to the examination of whatever material you are interested in, it is important to have a clear understanding of the physical principles applied in setting up a microscope. Also the correct utilization of the terms commonly applied in microscopy allows you to approach any problem from the correct departure point. Many of these basic principles will become clear once they are specifically applied to the operation of a microscope.

All microscopes make use of some form of electromagnetic radiation to directly or indirectly cause the formation of an image from a sample of interest. The image tells us about the structure and characteristics of a sample through interactions of light and matter. This book is dedicated to this principle: the application of visible light in some manner to form an image. How is an image created, and what components are important? Two main factors play a role: brightness and contrast. Whenever you make use of a digital camera to take a picture at some interesting location, the conditions may not be perfect or you may forget to select the correct camera setting. Hopefully the image is in focus and you can use any number of software packages to make adjustments to the image. One of the basic adjustments is setting brightness and contrast. The same adjustment applies to a microscope. Setting the brightness of the light source allows you to see an image of your sample and a contrast mechanism must be applied to make features visible.

2.1 REFRACTIVE INDEX

A physical property that is vitally important in microscopy, both in terms of the construction of a microscope as an optical system and the identification of a material, is the **refractive index**, defined as the ratio of the velocity of radiation in a vacuum relative to the velocity of the radiation in the material. When light moves from one medium into another (e.g., air into water), the light appears "bent." What causes this? The change in the electromagnetic environment experienced by the light that we already noted is modeled as an oscillating electrical field with positive (+) and negative (–) components.

A vacuum contains no matter, and the oscillations proceed unimpeded or unaffected by any other interactions. Air contains molecules of nitrogen, oxygen, water, and numerous other components. These molecules consist of electron clouds and nuclei that contain positively charged protons. Accordingly, the oscillating electrical wave experiences electromagnetic interactions (attraction and repulsion) that alter its velocity. Since air is not very dense, these interactions are not as numerous as they are in liquids and solids and thus the degree of refraction between air and water is very different from the refraction between a vacuum and the air, for example.

Snell's Law allows us to understand and measure refraction effects. As light moves from a medium with a relatively small refractive index (and travels faster) into a medium of relatively high refractive index (and travels slower), it will bend toward the normal (a line perpendicular to the incident surface) in the medium with a higher refractive index. Using Figure 2.1, Snell's law can be stated by the following equation:

$$\frac{n_1}{n_2} = \frac{\sin \theta_2}{\sin \theta_1}$$

(2-1)

This equation applies at any junction of two different media. A microscope involves many such junctions, such as between air and glass, sample and glass slide, and sample and mounting medium. The refraction is also exploited in the designs of optical components such as mirrors and lenses.

2.2 LENSES

We all know that a light microscope consists of a combination of lenses used to view a sample by means of direct visual observation of the projected image. It is

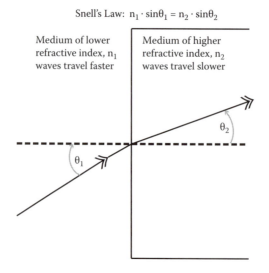

FIGURE 2.1   Geometrical description of Snell's law.

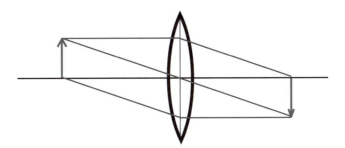

FIGURE 2.2   Construction of ray diagram to describe optical system.

important to understand the interaction of light with a lens to interpret these imag-es.* Practically speaking, a lens has a finite thickness. Some lenses, especially the front lens of a microscope objective, are significantly thick. To understand lens-light interactions, we consider the lens to be "thin" (like a line; also called a paraxial lens). Another way of thinking of a thin lens is that its thickness is small enough so as not to add to its **focal length** (the distance from the lens to the point where parallel rays entering the lens converge after exit). The **focal length** of a lens is determined by its radius of curvature and its refractive index. The thin lens equation is defined as

$$\frac{1}{f} = \frac{1}{a} + \frac{1}{b} \qquad\qquad (2\text{-}2)$$

where f = focal length; a = object distance (distance between object and lens); and b = image distance (distance between image and lens). The magnification (M) of an image formed by a lens is calculated as:

$$M = \frac{b}{a} \qquad\qquad (2\text{-}3)$$

To understand how light rays interact with a lens, we can make use of a geometric optical description. Typically we can use three rays in a ray diagram to understand a system. Figure 2.2 shows the basic construction of a ray diagram. A ray (chief or principal ray) that travels from a specimen (object) through the center of a lens does not deviate in its path. A ray that travels parallel from the object is refracted when passing through the lens such that the ray will pass through the **back focal point** (F'). A ray that travels from the object and passes through the **front focal point** (F) will be refracted when it passes through the lens such that the ray will be parallel to the optical axis. The distance between the lens and the front focal plane (or back focal plane) is called the focal length of the lens.† To study the effect of object dis-tance on a lens of given focal length, consider this series of scenarios:

---

* Remember that a sample can also act as a "lens."
† See scenario 1 (Figure 2.3).

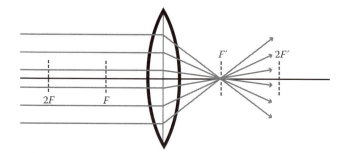

FIGURE 2.3   Ray diagram of scenario 1 (light source at infinity).

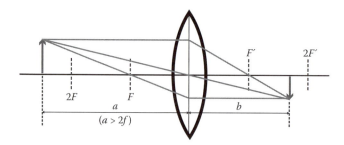

FIGURE 2.4   Ray diagram of scenario 2 ($a > 2f$).

Scenario 1 — Whenever a light source at an extreme distance (infinity) from a convex lens passes through that lens, the light will focus at a distance away from the lens known as the focal length of the lens (Figure 2.3). If the incident parallel rays of light do not enter the lens at 90°, then the focal point will be offset from the optical axis. This offset can be calculated as $y = f * tan\theta$.

Scenario 2 — If the object is greater than twice the focal length away from the lens ($a > 2f$), a real, inverted, and demagnified image is formed (Figure 2.3).* *Demagnified* implies that $M < 1$ (or that $b/a < 1$). When is an image real? A real image can be simply thought of as an image capable of projection onto a screen. If we place a screen at position P in Figure 2.4, the image of our object can be viewed on the screen and thus the image is real.

Scenario 3 — If we reduce the object distance so that the condition $a = 2f$ (Figure 2.5) is reached, we can see that b is also equal to $2f$.† In this case a real, inverted image is formed, but since $a = b = 2f$, $M = 1$.

Scenario 4 — If the distance is further reduced so that $f < a < 2f$, a real, inverted and magnified image is formed (Figure 2.6).

The above cases demonstrate that real and inverted images can be formed. The position of the object determines the magnification of the image. If the object distance is less than or equal to the focal length, virtual images are formed. A virtual

---

* Use the thin lens equation to calculate the value of *b*. Determine the magnification to understand the scenario.
† In Equation 2-2, substitute a with $2f$ and solve for *b*.

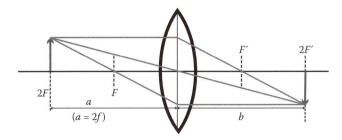

FIGURE 2.5   Ray diagram of scenario 3 ($a = 2f$).

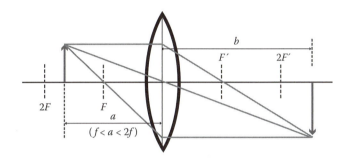

FIGURE 2.6   Ray diagram of scenario 4 ($f < a < 2f$).

image can be seen when you look through the lens toward the object. When is an image virtual? Simply stated, when it is not real. In other words, an image that cannot be projected onto a screen is a virtual image. As the object distance increases (from $a < f$) to the condition in which $a = f$, the image distance increases until it reaches infinity (at $a = f$). Under these conditions, no image exists to be projected onto a screen (opposite of scenario 1).

2.3   ABERRATIONS

So far we have considered our lenses to be perfect (focus always achieved at a single discrete point). Obviously this does not occur in reality. If a lens deviates from ideal, we can say that it has an **aberration**. An aberration is an effect that prohibits a lens from focusing at a specific point. The two most important aberrations of concern are chromatic and spherical. **Chromatic aberration** occurs when white light passing through a lens exhibits different points of focus, depending on the wavelength of the light. The refractive index of a lens is dependent upon the wavelength of light passing through the lens. This property is referred to as dispersion. For most transparent materials (e.g., glass) dispersion can be thought of as follows:

$$1 < n_{\lambda\text{-}red} < n_{\lambda\text{-}green} < n_{\lambda\text{-}blue} \tag{2-4}$$

This phenomenon is represented graphically in Figure 2.7. Dispersion can also cause colored fringes at the edges of objects in a specimen. The dispersion of a substance can be estimated by using Cauchy's dispersion formula:

$$n = A + \frac{B}{\lambda^2} + \frac{C}{\lambda^4} + \dots$$

(2-5)

where A, B, and C are constants for a specific compound. Thus, for white light passing through a given lens, a chromatic aberration will be seen with the shorter wavelengths focusing at a shorter distance than the longer wavelengths (Figure 2.8).

Because most lenses are not ideally thin (they have finite thicknesses), **spherical aberrations** occur. Rays that travel through the center of the lens and those traveling through the outer edges of the lens exhibit different path lengths through the lens and therefore are not all focused at exactly the same point. As a result, the image is not in the perfect image plane and a fuzziness of the image appears (Figure 2.9). Aspherical lenses are commonly used in cameras to reduce spherical aberrations.

There are four other types of spherical aberrations. The first three types result in image blur and are designated as **coma** (an off-axis light effect), **astigmatism** (focal points vary with planes, e.g., xz rays focus at a different point from xy rays), and **field curvature** (a positive lens forms a curved image; this effect can be corrected by special lens combinations). The fourth aberration type is **distortion**. If an optical system is constructed so that all other aberrations are corrected, the position of a specimen

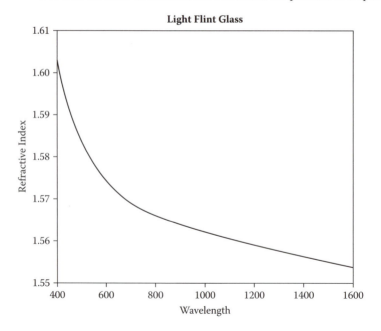

FIGURE 2.7   Dispersion of light flint glass.

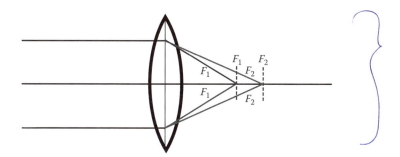

FIGURE 2.8 Ray diagram illustrating chromatic aberration. $F1$ represents blue light and $F2$ represents red light.

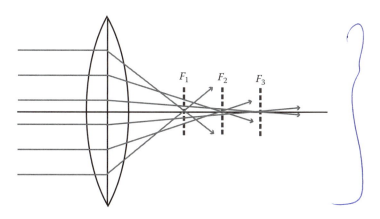

FIGURE 2.9 Ray diagram illustrating spherical aberration.

point may be off its predicted position. This is distortion of the image and can be stated as its relative deviation from its predicted position. The two types of distortion are positive (or pincushion) and negative (or barrel).

2.4 IMAGE FORMATION

We generally think of image formation as described by geometrical optics, but it can also be described another way. This second description is known as the **Abbe theory** of imaging (Figure 2.10). It was developed by Ernst Abbe who was hired by Carl Zeiss to design lenses for microscopes. This type of image formation can be best understood when **monochromatic** light is used. When white light passes through a prism, it is broken up into its component colors by a process known as diffraction. As mentioned in the previous chapter, diffraction is the interaction of light with a limiting edge, and this effect has consequences for microscope design and function. Diffraction can also be exploited as a means of breaking up white light into component wavelengths analogous to a prism, but with a different operating principle.

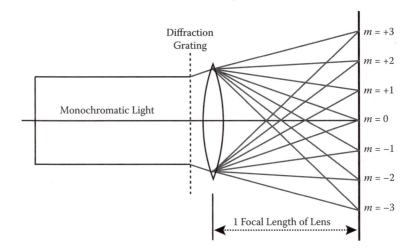

FIGURE 2.10 Image formation according to Abbe's theory.

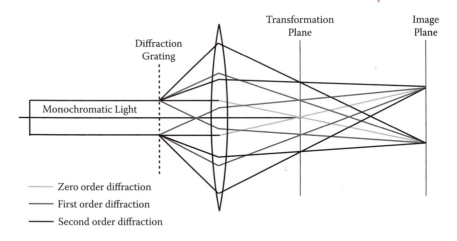

FIGURE 2.11 Diffraction of light by ruled diffraction grating.

This type of diffraction can be created using a **diffraction grating** (Figure 2.11). The classical grating is a series of transparent and opaque lines whose period is close to that of the wavelength of light. The edges cause the diffraction. A reflective grating has the profile of a saw blade and is typically found in **spectrometers**. The diffractive properties of the grating can be described by the grating equation:

$$\sin \beta_m = m \frac{\lambda}{\Lambda}$$

(2-6)

where β is the diffraction angle; λ is the wavelength of the incident light; Λ is the period of the grating (e.g., widths of lines); and m is an integer (diffraction order).

Consider monochromatic light passing through a grating. If a lens is placed beyond the grating, each of the orders of diffracted light can be determined in the back focal plane of the lens. The intensities of these focused spots of light are related to the grating. If the object that diffracts the light is not a grating or an object with a regular pattern, the light pattern found in the back focal plane of the lens will still describe the diffracting patterns within the object. A large object that varies slowly in its shading will result in little diffraction. Objects which are small or have fine details and sharp edges will produce significant diffraction.

If we place a lens more than one focal length from a grating (or object), a real image of the grating (or object) will be formed. Thus this image is formed by a recombination of the diffracted image that can be found in the back focal plane of the lens.

This concept of image formation is critical in microscopy and represents another way for light to interact with matter in a way we can literally see with a microscope. If a small object (with sharp edges) is placed on the stage of a microscope, it will act as a grating and higher orders of diffraction will occur. This is important in microscope design because the larger the collection angle, the better the resolution because the higher frequency objects (small and sharper) will occur further away from the optical axis. Only a little light is collected at the periphery of the lens, but this light makes a significant contribution to resolution of an image.

Clearly, the ability of a microscope to capture light after it has passed through a specimen determines its ability to resolve fine details within that specimen.* This light gathering ability is also of importance in telescopes. If the cone of light gathered by the objective is large, then smaller features in the specimen that cause more diffraction of light will be gathered than those features gathered by a smaller cone. This cone is thus defined by an angle that depends on two physical considerations: the diameter of the front lens of the objective and the distance of the objective from the specimen (Figure 2.12). This cone is called the **numerical aperture** (*NA*) of the objective. It is defined by the following equation:

$$NA = n \sin \theta \qquad (2\text{-}7)$$

The θ angle is half of the apex angle of the cone and *n* is the refractive index of the medium between the cover slip of the specimen and the objective. In most cases, *n* will be air, therefore *n* = 1.000; or in the case of oil immersion objectives, *n* = 1.515.[†]

---

[*] For a given grating and wavelength, the angle of diffraction is given by the order of diffraction. Because the high resolution information is retained in the higher orders, the larger the cone of light gathered by the objective of the microscope, the higher the resolution found in the image.

[†] What would the difference be in an image if immersion oil was placed between sample and objective? It may also be necessary to place oil between the front lens of the condenser and the sample to achieve the best resolution.

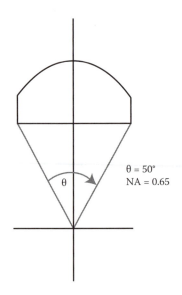

$\theta = 50°$
$NA = 0.65$

FIGURE 2.12   Numerical aperture (*NA*) of an objective.

2.5   RESOLUTION

An **Airy disk** (see Figure 2.13) is a diffraction pattern of a point source formed by a perfect optical system with a circular aperture. Compare this actual image to the idealized drawing in Figure 1.7 in the previous chapter. As in this figure, the pattern is a central circular bright area surrounding by alternating bright and dark concentric rings. Now, we can extend this knowledge to the microscope and evaluate its effects and consequences. In an Airy disk, the diameter of the first dark ring is given by

$$D = \frac{1.22\lambda}{NA}$$

$$(2\text{-}8)$$

The **Rayleigh criterion** is commonly used in microscopy to define resolution, and it derives directly from the Airy disk diffraction pattern. This criterion assumes that two points can be resolved if they are separated by the diameter of the first dark ring in the Airy disk:

$$D_{min} = \frac{0.61\lambda}{NA}$$

$$(2\text{-}9)$$

Resolution in an optical system such as a microscope is the ability of the system to differentiate small particles or structures. What is the smallest structure that can be resolved? This depends on resolution along with other factors, components of the microscope, and their integration into the optical train.

FIGURE 2.13   Image of Airy disk.

Located beneath the sample stage of a microscope is a critical component known as the **condenser** whose purpose is to provide a cone of light to illuminate the specimen. Among different types of condensers, the **Abbe condenser** is the most common because it is the simplest and least expensive. The Abbe type is most commonly constructed of two lenses and is uncorrected for spherical and chromatic aberrations. To obtain optimal performance, it is most preferable to have a condenser whose NA is greater than that of the highest NA objective.

In most circumstances, you will not change a microscope condenser very often. With a low power objective, the $NA_{condensor}$ will be greater than the $NA_{objective}$. When working with high power objectives and a simple Abbe condenser, you may find that the $NA_{condensor}$ will be less than the $NA_{objective}$. In both cases, different equations are required to determine the resolution of the system. In the case where the $NA_{condensor}$ is greater than the $NA_{objective}$,

$$d = \frac{0.61\lambda}{NA_{objective}} \tag{2-10}$$

where λ = wavelength of light used in the microscope and d = the smallest distance between two objects that can be resolved.

In the case where the $NA_{condensor}$ is less than the $NA_{objective}$,

$$d = \frac{1.22\lambda}{NA_{condensor} + NA_{objective}} \tag{2-11}$$

where λ = wavelength of light used in the microscope.

In Figure 2.14, the effect of the wavelength of light is given for an objective with 0.4 NA value. As can be seen from Equation 2-11, the resolution is directly proportional to the wavelength of the light. In Figure 2.15, the effect of numerical aperture

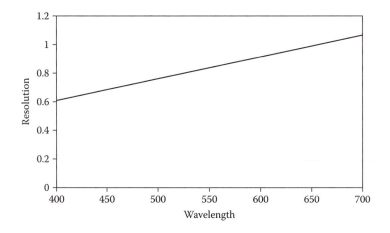

FIGURE 2.14   Resolution as function of wavelength.

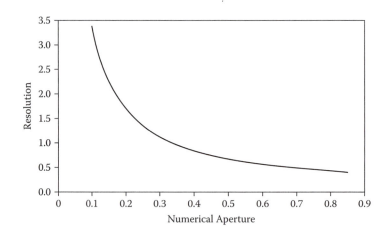

FIGURE 2.15   Resolution as function of numerical aperture.

on resolution is shown for a constant wavelength (555 nm). This graph displays the inverse relationship of numerical aperture and resolution. Figure 2.16 illustrates the same sample view with light filtered at different wavelengths. Notice the effects on the resolution or clarity of the images in each case. When setting a microscope up for Koehler illumination (KI; see Chapter 4), the field diaphragm is often fuzzy and surrounded by color fringes. This is typical of an Abbe condenser. An aplanatic or achromatic condenser may be used to improve the image.

2.6   DEPTH OF FIELD

Depth of field (*DOF*) in the object plane is the "thickness" of the sample in which all components are in focus:

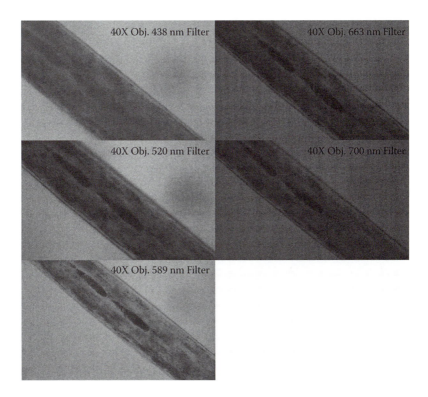

FIGURE 2.16 Human hair viewed at a number of wavelengths (notice effect of wavelength on resolution).

TABLE 2.1

Depth of Field and Resolution at Illuminating Wavelength of 555 nm

Magnification	NA	DOF (air)/μm	DOF (oil)/μm	Resolution/μm
5	0.10	55.500	84.083	3.386
10	0.25	8.880	13.453	1.354
20	0.40	3.469	5.255	0.846
40	0.65	1.314	1.990	0.521
60	0.85	0.768	1.164	0.398

$$DOF = \frac{n\lambda}{NA^2}$$

(2-12)

where n is the refractive index of the medium between the lens and the object. For green light (555 nm), the depths of field for selected NAs are listed in Table 2.1. DOF is useful for microscope set up and especially for photomicrography. If, for example, you have a sample that is 6 μm thick and you view it with a 40× objective with green light, the *DOF* is just over 1.3 μm. To improve the *DOF* using a 40× objective, you could close down the aperture diaphragm so that a *NA* of about 0.25 is reached to achieve a better depth of field, but with reduced resolution.

REFERENCES AND FURTHER READING

A. Abramowitz, *Microscope basics and beyond*, Olympus America, 2003.
D. Halliday and R. Resnick, *Fundamentals of physics*, John Wiley & Sons, 1981.
W. McCrone, L. McCrone, and J. Delly, *Polarized light microscopy*, McCrone Research Institute, 1995.
D. Murphy, *Fundamentals of light microscopy and electronic imaging*, Wiley-Liss, 2001.

QUESTIONS

1. Consider a situation similar to Scenario 1 in which light impinges on a paraxial lens at an angle of 35°. Construct a ray diagram indicating the focal point (give *x* and *y* coordinates) of this situation. Assume that you have a positive lens with a focal length of 45 mm.
2. For a positive lens with focal lengths of 10, 15, 20, and 30 mm, calculate the value for *b* given that *a* values are 35, 60, 90, and 120 mm, respectively. For each situation, calculate the magnification factor.

EXERCISES

Exercise 2.1: Airy disk

Materials needed:

    Laser
    Small aperture (about 25 μm)
    Imaging screen
    Camera

In a dark room, project a laser beam onto the screen. Introduce the aperture into the laser beam so that it is correctly aligned. After alignment is achieved, move the screen toward the aperture so that the disk is in focus on the screen. After it is focused, take an image of the screen. The image can be processed in ImageJ (see Chapter 6) and the relevant channel extracted based on laser color (red or green). Adjust the image for brightness and contrast. The diffraction fringes of the Airy disk will now be visible. Compare your results to Exercise 1.1

Exercise 2.2: Evaluation of objectives

Materials needed:

Microscope
Details of all objectives
Specimens

Use Equation 2-6 to determine half of the apex angle of the cone. Multiply by two to estimate the actual size of the cone. Using the half angle, and after measuring the width of each objective lens, estimate the working distance (WD) of each objective. Look for any relationships among NA, WD, and magnification.

The next step is to determine the resolution of each objective. Establish the NA of the condenser (if you cannot find it, assume it to be 0.85). Now use Equation 2-9 or 2-10 to determine the resolution for each objective and the condenser. Using Equation 2-11, calculate the DOF for each objective using the following wavelengths: 475 nm (blue), 510 nm (green), 570 nm (yellow), 590 nm (orange), and 650 nm (red).

3 Types of Microscopes

A microscope can be made from basic components as illustrated in the previous chapter. For a microscope to operate in an effective manner, these components must be combined in a mechanically stable system (a stand; see Figure 3.1). Remember that microscopy involves magnification, resolution, and contrast. This microscope stand allows introduction of a number of different contrast techniques to improve the versatility of the microscope.

This chapter is divided into two sections. The first covers the main components of a microscope and the second describes the main types of microscopes encountered in laboratories. In effect, this chapter is a tour of microscope components starting from the objectives and working down to the lamp that illuminates the optical train. These components provide magnification and define resolution while contrast can be affected by different microscope techniques. We will introduce some of these methods and describe others, such as bright field and polarizing light methods, in much greater detail in subsequent chapters.

3.1 MAIN COMPONENTS

3.1.1 OBJECTIVES

The most important optical component of a microscope is the **objective lens** (Figure 3.2), which is composed of a number of individual lenses combined within the barrel of the objective. The multiple lenses are placed in a specific arrangement to reduce both spherical and chromatic aberrations and to allow for a flat field of view. Most objectives are designed to require the use of a cover slip over a sample. A piece of flat glass will affect the optical path of a light ray, thus the absence of a cover slip, when required, will result in degradation of the image of the specimen. A microscope may also include a **nosepiece** into which a number of objectives can be mounted (see Figure 3.3). The three main types of objectives are achromats, fluorites, and apochromats.

3.1.1.1 Achromats

These are the cheapest objectives and have few corrections. They are corrected for chromatic aberration to ensure that red and blue lights are brought to focus at a common point. **Achromats** are corrected for spherical aberrations with green light. They are best used when the light source utilizes a green filter and images are collected in grayscale.

3.1.1.2 Fluorites

These objectives are also known as **semi-apochromats**. The chromatic correction is similar to the correction of achromats but the focal point of the red and blue light

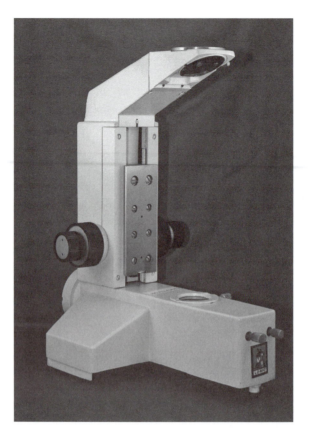

FIGURE 3.1 A microscope stand must be stable, accommodate all attachments, and provide a clear optical pathway.

is closer to that of the green light. Fluorites are also corrected for two wavelengths (green and blue) to control spherical aberrations. These objectives work well for color photomicrography.

3.1.1.3 Apochromats

Apochromatic objectives are chromatically corrected for two wavelengths in the blue region as well as for green and red. They are spherically corrected for three wavelengths. They generally contain greater numerical apertures than the other objectives. These are the highest quality objectives and are ideally suited to color imaging.

3.1.2 EYEPIECES

When selecting an eyepiece, two requirements must be met: field of view (FOV) and magnification. The maximum magnification optical system should be 500 to 1000 times that of the numerical aperture. The eyepiece should be chosen not to exceed

FIGURE 3.2 100× 1.25 *NA* achromat objective.

FIGURE 3.3 Nosepiece holding five objectives. Note that adjusting screws allow centering of objective in polarized light microscopy.

this requirement and should be lower in magnification than the objective. The FOV and magnification values are normally inscribed on eyepieces.

3.1.2.1   Huygenian (Negative) Eyepieces

The Huygenian is a simple eyepiece consisting of an upper eye lens and a lower field lens (see Figure 3.4a). Each lens is **plano-convex,** with plane sides on the eye side. A fixed diaphragm is set between the lenses. The Huygenian is the most economical type and lacks correction for chromatic aberrations.

3.1.2.2   Ramsden (Positive) Eyepieces

A doublet lens is employed for both eye and field lenses (see Figure 3.4b). The fixed diaphragm is found below the field lens.

3.1.3   CONDENSERS

The function of the condenser is to provide a cone of light to illuminate a specimen. The **substage condenser** is placed between the light source and the stage. A typical Abbe condenser has a numerical aperture up to about 1.25. The Abbe condenser has no correction for spherical or chromatic aberrations. The cone of light passes through the specimen and reverses itself to fill the front lens of the objective. The aperture diaphragm controls the angular size of the cone. The condenser must be focused correctly when a microscope is set up for use. This will be discussed in the section on Kohler Illumination (Chapter 4).

When choosing a condenser for a microscope, it is important that the numerical aperture of the condenser be greater than or equal to the numerical aperture of the largest objective. This will ensure that the angular size of the cone of light leaving the condenser matches the cone of light required to provide the greatest efficiency

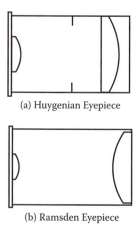

(a) Huygenian Eyepiece

(b) Ramsden Eyepiece

FIGURE 3.4   (a) Huygenian and (b) Ramsden eyepieces.

FIGURE 3.5 A 0.85 *NA* condenser with turret and lever (to adjust size of condenser aperture).

for the objective. If a condenser has a numerical aperture of 0.85 (for example), the device is an air condenser and no oil immersion is required to achieve greatest efficiency (see Figures 3.5 and 3.6). If a microscope has an objective with an *NA* value of 1.2, this condenser will not be able to provide a cone of light sufficient to cover the front lens of the objective.

When an Abbe condenser is used and the field diaphragm is imaged in the condenser, the leaves of the diaphragm will not be sharp and will exhibit color fringes on the edges. This is acceptable for routine use, but for good photomicrography and analytical procedures, an **aplanatic–achromatic** condenser is required. These condensers have multiple lenses to correct for aberrations.

3.1.4 STAGE

The stage provides specimen support. For most biological microscopes, a stage that can move on *x* and *y* axes is sufficient. For applications requiring angular measurements and polarized light microscopy (PLM), a rotating stage (see Figure 3.7) is an absolute necessity. The slide can be secured using clips or a mechanical stage to maneuver it around to view items of interest (see Figure 3.8).

FIGURE 3.6   A 0.85 *NA* condenser from below with leaves of condenser aperture visible.

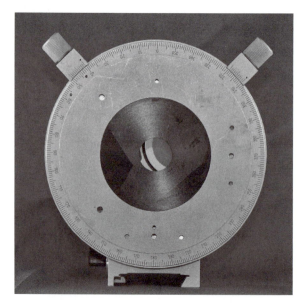

FIGURE 3.7   Rotatable stage for PLM (note 1-degree graduations plus Vernier scale).

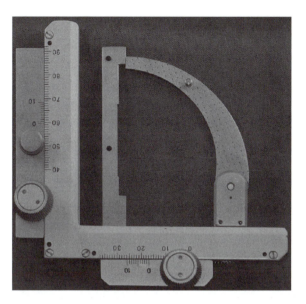

FIGURE 3.8 Mechanical stage for positioning slides. Approximate position of item of interest can be recovered by using an England finder.

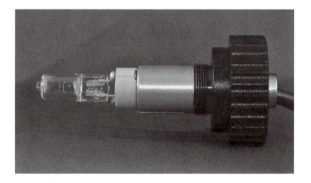

FIGURE 3.9 Tungsten lamp in holder.

3.1.5 LIGHT SOURCES

Most microscopes make use of tungsten lamps of 6 to 12 volts (see Figure 3.9). The color temperatures of these lamps vary between 2700 and 3200 K. The color temperatures can be explained by a **black body** radiator.* We have all heard the term *white hot* and thus intuitively understand the concept of heat associated with a color. This color can also be associated with a temperature. As we heat a black body radiator, it will move through a number of colors. Each color is then related to a temperature

---

* A black body is an object that absorbs all radiation incident upon it.

FIGURE 3.10 Lamp holder (see Figure 3.9) inserted into lamp housing (on right). Threading on the lamp holder allows adjustment of the filament on the *x* axis. The rod just above allows adjustment along the *y* axis. Looking down the optical axis, one sees two levers. The front lever adjusts the focus of the filament and the second lever adjusts the field diaphragm.

on the Kelvin (K) scale. Thus we use the color temperature of a light to achieve a consistent description.

The illumination set up requires a collector lens to project the image of the lamp and a field diaphragm to control the size of the illuminated field (a variable iris diaphragm). The ability to adjust the position of the filament (lamp) is clearly advantageous to achieve even illumination (see Figure 3.10). When using a filter near the field diaphragm, it is very important that such filters be very clean. Any particle on a filter will produce a slightly defocused image in the specimen plane Why? See Chapter 4 on Kohler Illumination. Another solution to this problem would be to place the filters just below the substage condenser.

Figure 3.11 depicts the function of the collector lens on the lamp holder. The lamp holder has been removed and the position of the filament adjusted. The collector lens is used to focus the filament on a distant object.

3.2 TYPES OF MICROSCOPES

Remember that to image a specimen, contrast of the image is required. A number of chemical and physical techniques can be applied to improve contrast. The optical set up required to achieve each of these contrast techniques is discussed below.

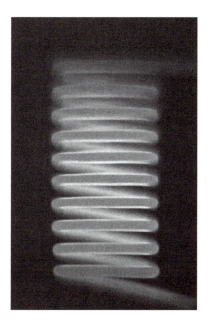

FIGURE 3.11   Projected image of tungsten filament. A similar image is obtained during set up of Kohler Illumination upon insertion of Bertrand lens.

3.2.1   Bright Field Microscopy

This is the basic microscope set up. The main components are an eyepiece, objective, condenser, and light source. Bright field microscopy can be used for basic observation and image recording. This set up is described in detail in Chapter 4.

3.2.2   Dark Field Microscopy

Some sample types are not visible or are only barely visible under bright field conditions. The visibility of such specimens can be improved by dark field illumination. A dark field stop is placed in the light path between the light source and the substage condenser (see Figure 3.12). The size of this stop will limit any light that, under bright field conditions, would enter the objective. The light diffracted or scattered by the sample specimen will then enter the objective and result in image formation against a dark background. In this set up, the aperture diaphragm should be completely open.

3.2.3   Phase Contrast Microscopy

Amplitude objects are viewed as having inherent high contrast. The changes in amplitude result in the contrast. Some specimens (especially biological ones) do not produce the inherent contrast, and simply cause a phase shift in the rays passing through

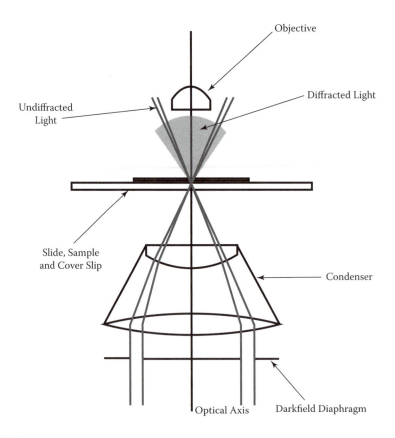

FIGURE 3.12 Optical set up for dark field microscopy.

them. These are known as phase objects. A phase contrast microscope converts these changes in phase into contrast (similar to chemical staining of a sample).

When light passes through a phase object, most of it will undergo no diffraction (S wave). A relatively small portion will be diffracted (D wave). Both waves enter the objective and undergo destructive interference to form a resultant P wave. If the P wave is not significantly different from the S wave, no image will be seen. When imaging a phase object, it is important to note the phase relationship between the S and D waves, and maintain that phase relationship between the sample and the image.

When we look at the S and D waves in the back focal plane of the objective, it should be noted that the phase difference between them is a quarter of a wavelength. When the S and D waves combine to form the P wave, the phase difference between the S and P waves is reduced to only a twentieth of a wavelength. Thus the P wave approximates the S wave and therefore the image shows no contrast.

A phase contrast microscope is configured by replacing the condenser diaphragm with a **condenser annulus** so that the sample is illuminated only with light that passes through the annulus (see Figure 3.13). Since we will also set up our microscope to take advantage of Kohler Illumination (Chapter 4), the phase contrast

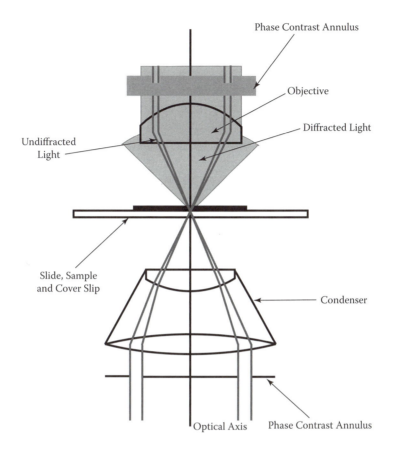

FIGURE 3.13   Optical set up for phase contrast microscopy.

technique also takes advantage of the fact that the undiffracted light (S wave) passing through the condenser annulus will be focused as a bright ring in the back focal plane of the objective. The diffracted light will be distributed in the back focal plane of the objective. This differentiation of these two waves in space allows us to operate on them independently. This is achieved by "inserting" a plate near the back focal plane to adjust the S wave. For positive phase contrast, the plate also has an annulus etched into it to advance the phase of the S wave by a quarter of a wavelength. This annulus is coated with a film to significantly reduce the amplitude of the wave. When the S and D waves combine to form the P wave, the result is that the P wave is not equal to the S wave, which, in turn, results in contrast in the image.

3.2.4   POLARIZED LIGHT MICROSCOPY

Polarized light microscopy can be used to introduce contrast into an image, but is most useful as an analytical technique. To highlight its usefulness, we have included a complete chapter on its applications (see Chapter 7).

3.2.5   OTHER METHODS

A variety of other methods available for contrast enhancement include differential interference contrast microscopy and Hoffman modulation contrast microscopy. For more information on these techniques, consult the reference section.

REFERENCES AND FURTHER READING

A. Abramowitz, *Microscope basics and beyond*, Olympus America, 2003.
R. de Levie, *Advanced Excel for scientific data analysis*, Oxford, 2004.
J. Delly, *Photography through the microscope*, Eastman Kodak, 1988.
R. Ditchburn, *Light*, Blackie & Son, 1963.
R. Fischer, T.-G. B and P. Yoder, *Optical system design*, SPIE Press, 2008.
D. Halliday and R. Resnick, *Fundamentals of physics*, John Wiley & Sons, 1981.
W. McCrone, L. McCrone, and J. Delly, *Polarized light microscopy*, McCrone Research Institute, 1995.
D. Murphy, *Fundamentals of light microscopy and electronic imaging*, Wiley-Liss, 2001.
D. O'Shea, *Elements of modern optical design*, John Wiley & Sons, 1985.

QUESTIONS

1. Why is the objective lens the most important optical component of a microscope?
2. Which components of a microscope provide the necessary magnification, resolution, and contrast, and why?
3. What role does dispersion play in the quality of an objective lens?
4. How can a lens be corrected for spherical aberrations?
5. Why should the numerical aperture of the condenser be greater than or equal to the numerical aperture of the largest objective?

EXERCISES

Exercise 3.1: Phase shifts of two rays

Material needed:

Spreadsheet application such as Microsoft Excel® or OpenOffice.org Calc

Use a spreadsheet to represent the effects of phase shifts of two rays. Assume that the wavelength of a light is 2π ($\approx 360°$). This assumption makes it easier to perform the calculations, but you could have chosen a wavelength of 555 nm. Use the sine function to calculate the amplitude of the wave. Remember that the parameter of the function is in radians and not degrees. If you prefer to use degrees, then the formula should be constructed as follows: =sin[radians(value in degrees)]. Perform this calculation for two waves. For the second wave, introduce a shift between the waves. To calculate the shift, select a cell for the shift (in degrees) as a constant. Adjust the calculation for the amplitude of the wave as =sin[radians(value in degrees + shift)]. Using a constant cell for the shift will allow you to evaluate the effects of different shifts. Plot both waves using a scatterplot type graph. Remember that your first column should be a list of degrees and the next two columns should represent calculations of the amplitude for each wave.

Exercise 3.2: Phase contrast

Material needed:

Spreadsheet application such as Microsoft Excel® or OpenOffice.org Calc

Use a spreadsheet to represent the effects of phase contrast. First, adapt the spreadsheet from Exercise 3.1. Use the first wave column to represent the S wave and the second wave column to represent the D (shifted) wave. We know that the shift between the S and D waves is about one quarter of a wavelength (360°/4 = 90°). Use 90° as the shift between the waves. We also know that the intensity of the D wave is significantly less than that of the S wave. To recalculate the D wave, adjust the formula: =sin[radians(angle + shift)]*factor. Use a cell for the factor so that you can vary it to observe the effects. Now calculate the resultant P wave. Create a third column and add the values of the S and D waves. In your scatterplot graph, note that the P wave is extremely close to the S wave. What does this mean?

To evaluate the use of a phase plate, we must make some adjustments. We know that a phase plate for positive phase contrast has an annulus etched on it to advance the phase of the S wave by a quarter of a wavelength. This annulus is also coated with a film to significantly reduce the amplitude of the wave.

We now need to create a new value for the S wave. Create a new column labeled S wave PC. The formula will now take the form of =sin[radians(angle + 90°)]*factor pc. Create a new column for the phase contrast. Compare the S wave pc to the P wave pc. What will happen to the image now?

Use values for the phase contrast factor (intensity reduction) of 0.7, 0.5, etc. What happens if the phase shift on the D wave is slightly different from the shift (90°) on the phase plate? Create a new graph and compare the brightfield scenario with the phase contrast scenario.

4 Kohler Illumination

The main objectives of microscope set up are magnification, resolution, and contrast. The first two depend on the components used in the microscope. The latter depends on how the microscope is adjusted. To achieve optimum contrast, it is essential that the illumination be correct. In turn, contrast dictates how well you can see samples and photograph them. There are a number of ways of setting up microscope illumination. Initially **Nelsonian** or **critical illumination** was used, but today the most widely applied technique is **Kohler Illumination**.

Most microscopes today make use of built-in lamps. A current is passed through the lamp filament and the filament starts to radiate light. As noted in Chapter 1, the filament serves as a non-homogeneous source of light for the microscope and covers a range of visible light wavelengths at various intensities. When imaging a sample in a microscope, even illumination of the background is critical. The optical set up should provide homogeneous light at the highest possible intensity.

This chapter includes a description of the set up for Kohler Illumination. Although the set up steps may appear tedious at first, its implicit benefits will be appreciated after extended use of the technique, especially when you want to produce film or digital images.

4.1 DESCRIPTION

Kohler Illumination requires evaluation of two sets of rays: the **illuminating rays** and the **image forming rays**. Although we can consider them independently, they obviously do not exist in isolation. Understanding of the rays provides a simplified understanding of the optical set up. However, while points are used to describe the light paths, the whole of the filament is used in determining the final image.

The next concept essential in Kohler Illumination is the use of **conjugate planes**—a set of planes in which a particular point is in focus in each individual plane. Each set of rays has its own set of conjugate planes. Two main tasks are necessary for achieving Kohler Illumination. First, the image of the field diaphragm must be focused in the specimen plane by using the focus of the substage condenser. Second, the image of the filament must be focused in the plane of the condenser diaphragm using the focus lens on the lamp.

The general approach to achieving Kohler Illumination involves several steps. First, place a sample slide on the stage and achieve focus using a 10× objective. This is done simply to define a starting point for the set up. The field diaphragm is closed down and the position of the substage condenser is adjusted until the field diaphragm is in focus. The image of the field diaphragm must be adjusted for both position and size. The method for adjusting size is determined by the particular microscope, but is achieved commonly via a set of adjusting rods found on the substage condenser assembly. **Diffraction colors** may be seen (usually with an Abbe condenser and a higher power objective) at the edges of the leaves of the diaphragm. The focus of the substage condenser should be adjusted to minimize the diffraction. To do so, adjust to maintain the green diffraction color that lies in the middle of the visible spectrum. After the field diaphragm is centered, adjust the size of the diaphragm so that it extends just beyond the field of view. If the position of the field diaphragm has been correctly adjusted, all the leaves of the diaphragm should leave the field of view at the same time when the size of the aperture is increased.

Focusing the field diaphragm is relatively simple because it involves the same set of conjugate planes as the specimen. When Kohler Illumination is achieved, both the specimen and the field diaphragm are simultaneously in focus. It can also be determined by the formation of the image of the specimen on the retina (or film plane of a camera) that the retina is also conjugate to the field diaphragm and the specimen, leading to the conclusion that there must be another conjugate plane between the specimen and the retina. This is found in the lower part of the eyepiece and is known as the intermediate image plane. Typically, there is a shelf (fixed diaphragm) inside the eyepiece. **Thus, the four conjugate planes in the image forming ray set are the field diaphragm, specimen, intermediate image plane, and retina** (see Figure 4.1). Remember that the lens of the eye serves as part of the optical set up.

After the focus of the field diaphragm has been achieved, the lamp filament must be focused in the plane of the aperture diaphragm. A different strategy is required to meet this requirement. Since the lamp filament is a conjugate plane in the illuminating ray path, we must devise a method to view this set of conjugate planes. This can be done by removing an eyepiece (with or without inserting a pinhole into the eyepiece socket) or by inserting an additional lens into the optical path. The insertion of this lens allows the illuminating path conjugate planes to be in focus at the retina.

When this set up is achieved, the focusing element on the lamp may be adjusted so that the filament is focused. It may also be necessary to adjust the position of the filament so that it fills the image. A filament off to one side may result in unbalanced illumination. Since one of the fundamental outcomes of the Kohler technique is even illumination, unbalanced illumination defeats the purpose of the set up.

This can be explained in terms of the conjugate planes of the illuminating ray path. After the filament and the aperture diaphragm, the next conjugate plane is found at the back focal plane of the objective. When viewing ray diagrams, you must also remember the concept from Chapter 2 that each simple lens represents a complex lens component. After passing through the objective, the illuminating ray path is brought to focus at the **eyepoint** (about 10 mm above the eyepiece). This represents the final conjugate plane of the illuminating path. **Thus, the four conjugate planes in the illuminating ray set are the filament, condenser diaphragm, back focal plane of the objective, and eyepoint.** Although this focusing can be done using a **Bertrand lens**, why does the action of simply removing an eyepiece also work? Without an eyepiece, the illuminating rays are focused in the back focal plane of the

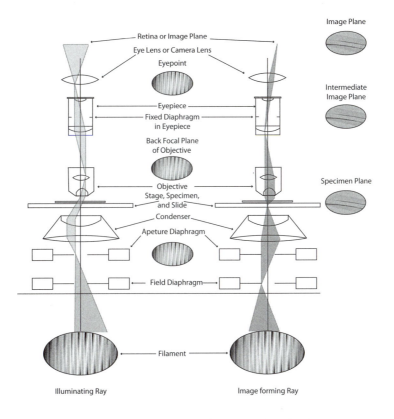

Image Plane

Intermediate
Image Plane

Specimen Plane

Retina or Image Plane
Eye Lens or Camera Lens
Eyepoint

Eyepiece
Fixed Diaphragm
in Eyepiece

Back Focal Plane
of Objective

Objective
Stage, Specimen,
and Slide
Condenser
Apeture Diaphragm

Field Diaphragm

Filament

Illuminating Ray   Image forming Ray

FIGURE 4.1 Conjugate planes in microscope after Kohler Illumination set up.

objective. You can verify this by opening and closing the substage condenser and noting the effects.

4.2 SET UP AND TECHNIQUE

The process of setting up Kohler Illumination will be described photographically. Obviously all microscopes are different, but the procedure described here can be used as a broad outline. Usually, the manufacturer of a microscope can provide detailed instructions for a particular instrument; check the manual or the company's website as a starting point. If the Kohler set up is performed in conjunction with a review of your microscope manual, you will be able to set up your microscope correctly. You may also want to refer to Figure 4.1 as you work through this process to remind yourself how each adjustment affects the image you see.

Step 1 — Place a prepared slide onto the stage. Turn on the lamp and open all apertures on the microscope (Figure 4.2).

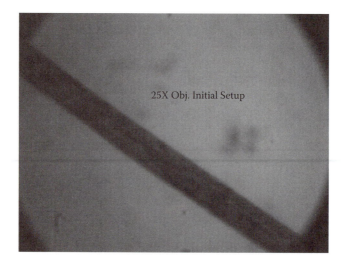

FIGURE 4.2   Step 1: A sample (hair) is placed on the stage and light is passed through the microscope.

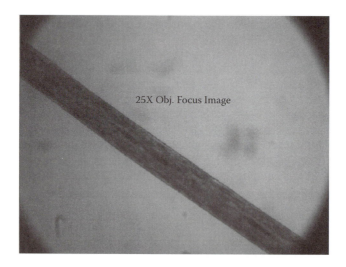

FIGURE 4.3   Step 2: Sample is brought into focus.

Step 2 — Bring the image into focus by adjusting the coarse and fine settings. Set the **interpupillary distance** and the **diopter** correction of the eyepieces. If your microscope has one eyepiece with a diopter adjustment, blank off that eyepiece. Using only the eyepiece without a diopter adjustment, adjust the fine focus until a small speck is well focused (Figure 4.3). Then close off that eyepiece and adjust the diopter setting of the other eyepiece until the speck is in focus. The diopter adjustment should now be set correctly.

Step 3 — The next series of steps will adjust the light source and field diaphragm. Close down the field diaphragm (Figure 4.4).

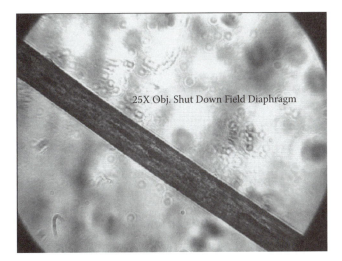

FIGURE 4.4 Step 3: Field diaphragm is closed down.

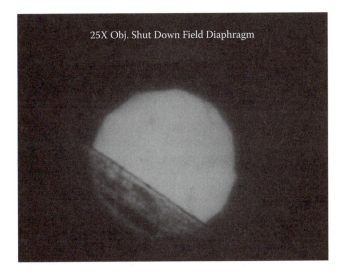

FIGURE 4.5 Step 4: Field diaphragm is focused by adjusting condenser position.

Step 4 — At this point, the image viewed may be variable. See Figure 4.5. We want to ensure that the field diaphragm is in focus in the plane of the sample, and then focus on the leaves of the field diaphragm by adjusting the position of the condenser relative to the sample slide. The condenser is on a rack-and-pinion drive. Turn the knob to drive the condenser up and down. Stop when the leaves of the field diaphragm are clearly focused. The quality of focus will depend on your microscope stand, objects, and condenser. You may see some diffraction fringes around the edges of the leaves. They can be minimized by using a filter.

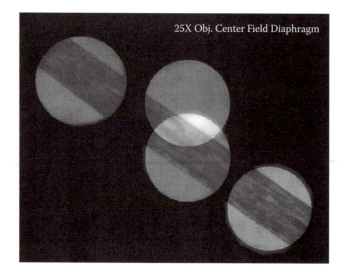

FIGURE 4.6   Step 5: Field diaphragm is adjusted to ensure that it is centered. Image shows position of the field diaphragm after a few adjustments.

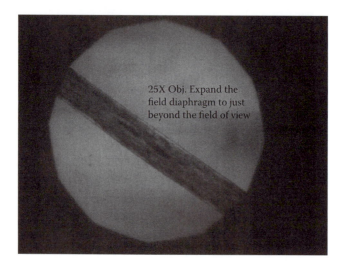

FIGURE 4.7   Step 6: Field diaphragm is expanded until it is just beyond the field of view.

Step 5 — We now want to adjust the position of the field diaphragm so that it is centered in the field of view. See Figure 4.6. The knobs and/or levers used to do so depend on the model of microscope. They may be near the condenser or elsewhere. Refer to your user manual to determine which knobs or levers should be used. Be aware the levers designed to center items in the condenser turret will not help. To check whether the field diaphragm is centered, increase its size. If it is centered, it should touch the edges of the field of view simultaneously. If it is not centered, close it down a bit and re-adjust.

FIGURE 4.8 Step 7: Bertrand lens is inserted into optical path and focused on image of filament.

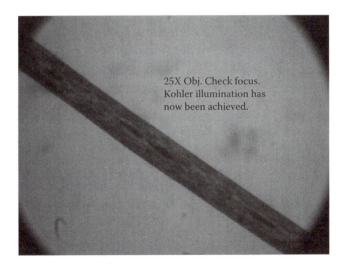

FIGURE 4.9 Step 8: Remove Bertrand lens and ensure that the image of the sample is still in focus. Kohler Illumination has now been achieved.

Step 6 — Expand the field diaphragm until it is just beyond the field of view (Figure 4.7).

Step 7 — We now need to adjust and focus the filament. The Bertrand lens is inserted into the optical path and focused on the image of the filament. If you do not have a Bertrand lens, remove one eyepiece and view down the tube (some microscopes have a pinhole or focusing tube for this purpose). The image of the filament is focused using the adjustment of the collector lens on the lamp housing. The position

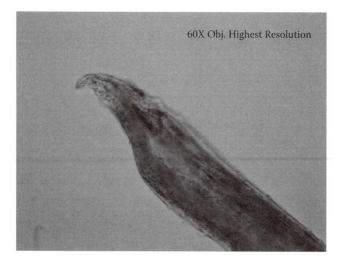

FIGURE 4.10 High resolution image of a human hair with a 60× objective.

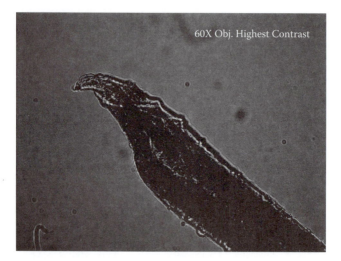

FIGURE 4.11 High contrast image of a human hair with a 60× objective.

of the filament must be adjusted to ensure that it fills the field of view. The lamp housing will include features that allow adjustment of the position of the filament in both the *x* and *y* axes. See Figure 4.8.

Step 8 — Remove the Bertrand lens (or telescope or pinhole); re-insert the eye-piece. The image should be in focus and Kohler Illumination has been achieved (Figure 4.9). **Remember that this set up must be followed whenever an objective is changed.** After you have performed the procedure a few times, it becomes second nature.

Step 9 — When viewing a sample, the optical set up must be adjusted for contrast and resolution. This is achieved by adjusting the size of the aperture diaphragm.

The highest resolution and highest contrast images for the same sample are shown in Figures 4.10 and 4.11. The optimal setting is somewhere between these extremes, depending on your needs.

REFERENCES AND FURTHER READING

A. Abramowitz, *Microscope basics and beyond*, Olympus America, 2003.
J. Delly, *Photography through the microscope*, Eastman Kodak, 1988.
R. Ditchburn, *Light*, Blackie & Son, 1963.
D. Halliday and R. Resnick, *Fundamentals of physics*, John Wiley & Sons, 1981.
W. McCrone, L. McCrone, and J. Delly, *Polarized light microscopy*, McCrone Research Institute, 1995.
D. Murphy, *Fundamentals of light microscopy and electronic imaging*, Wiley-Liss, 2001.

QUESTIONS

1. When using a filter close to the field diaphragm, it is very important that such filters be very clean. Why?
2. The collector lens on the lamp holder can be used to focus the filament. Why do we want to do this?
3. We can view the conjugate planes of the illuminating rays by inserting a Bertrand lens. This can also be achieved by removing an eyepiece. Why does this also work?
4. After you have set up Kohler Illumination, you can reduce the brightness of the image by closing down the field diaphragm or defocusing the condenser. Is this the correct technique?
5. What are the advantages of Kohler Illumination?
6. The numerical aperture of a microscope is given as the average of the *NA* of the objective and the condenser. If the *NA* of the condenser is 1.25, when would it be appropriate to use this equation?

EXERCISES

Exercise 4.1: Evaluation of conjugate planes under Kohler Illumination

Materials needed:

Microscope set up for Kohler Illumination
Sample with little detail
Filter markings or graticule* (referred to as "the filter" in instructions below)

After you have set up the microscope according to Kohler Illumination technique, place the filter into the optical path at the positions of the conjugate planes. Start with the image forming rays, then repeat for the illuminating rays (remember you will have to use a Bertrand lens or remove the eyepiece). Once you have

---

* You can substitute a piece of acetate sheet with small markings (about the size of a typical specimen). CAUTION: Acetate will melt if it is placed too close to a lamp.

completed evaluating insertion of the filter at the conjugate planes, place the filter slightly off the conjugate plane (higher or lower, as practical). Evaluate the effect. How can you use this knowledge to your benefit?

Exercise 4.2: Evaluating contrast and resolution of objective*

Materials needed:

> Microscope set up for Kohler Illumination
> Diatom sample or sample with many fine details
> Camera on microscope

Adjust all settings on your microscope (field diaphragm, aperture diaphragm, condenser focus) so that they are in an arbitrary position. Place your sample on the stage and achieve the best focus. Record the image. Now adjust the condenser diaphragm until the best resolution is achieved (open) and record the image. Set up Kohler Illumination with your aperture diaphragm set to achieve the best trade-off between resolution and contrast. Open the condenser aperture to achieve highest resolution and record the image. Compare the images in terms of contrast and resolution. Repeat the exercise for all your other objectives.

---

* You may need to read Chapter 8 before attempting this exercise.

5 Sample Preparation, Manipulation, and Micrometry

This chapter will discuss the fundamentals of sample handling: how to select the correct sample or subsample, what material (if any) to use for mounting, and how to measure sample components such as fibers or particles. We will focus on basics and general principles; the procedures selected for any task will be application-specific. What is *always* true is that the appropriateness of the sample—how you select and prepare it—dictates the quality and reliability of the results.

5.1 PREPARING SAMPLES: BASIC PRINCIPLES

We will look at three steps in the sample preparation process: (1) obtaining a representative microsample from a larger bulk sample; (2) methods of manipulation and preparation; and (3) mounting and preserving samples. Before beginning sample preparation, you should consider a few pertinent questions such as

- What information *must* I obtain from this sample?
- What information would I like to obtain?
- How much of the sample can I use?
- Is the sample heterogeneous or homogeneous?
- Is the mounting temporary or must I preserve my slides?
- What other analytical methods will be performed?
- Will the refractive index of the sample compared to that of the mounting media be an issue (for example, with glass)?

The answers to these questions dictate how you obtain and prepare samples. Keep in mind that the answers and even the questions are likely to be different for every analysis. The universe of samples that a microscopist may encounter is endless and it is impossible to adequately describe even a general approach to sampling beyond stressing the importance of obtaining microscopic samples that accurately represent the composition of a **bulk sample**, whatever that may be.

Obtaining a **representative sample** for microscopy may be a challenge due to the small scale of the analysis. To illustrate, consider about 100 mg of an unknown white powder that you are to characterize microscopically. As a general rule, you should never consume an entire sample unless you have absolutely no alternative. The size of the white powder sample is adequate for microscopic analysis so assume we can work with up to half of the sample or about 50 mg. Which half? A sample cannot be divided until it is determined to be homogeneous. One method among several is observation under a stereomicroscope. This will also provide information about the number of components in the mixture because it will reveal different crystal morphologies. You can also mix the bulk sample and visually determine whether, based on particle distribution, it appears to be well mixed. At that point, you can collect a subsample and continue.

With a larger powder sample such as a box of laundry detergent, a bag of sugar, or a kilogram of cocaine, the sampling problem becomes more difficult. The guiding principle in any such task is to obtain random and representative samples that may constitute several individual samples or some form of composite sample. The key issue is that the sampling plan must be reasonable and defensible. While a detailed discussion of sampling theory exceeds the scope of this book, many good references include Chen et al. 2008; Esbensen et al. 2007; Garfield 1989; Gy 1995; Holm-Nielsen, Dahl, and Esbensen 2006; Paakkunainen, Reinikainen, and Minkkinen 2007; Petersen and Esbensen 2005; Petersen, Minkkinen, and Esbensen 2005; and Springer and McClure 1988.

While mixing to obtain apparent homogeneous composition should be undertaken, grinding may not be a good choice for microscopic work because the morphology of the crystals in the mixture would likely be lost or compromised. For mixing without grinding, one common method involves **coning and quartering** in which a sample is placed on a clean surface such as glass and gathered into a cone-shaped pile. The cone is divided into four roughly equal piles and each pile is thoroughly mixed individually. The cone is reconstituted and the procedure repeated several times.

After a reasonable and defensible sampling plan has been designed and a representative sample or samples obtained, sample preparation can be considered. In the simplest case, samples can be transferred to a slide and observed with or without a cover slip, although as we learned in previous chapters, the presence or absence of a cover slip can affect focusing and illumination. Slide cleanliness is critical. Avoid cleaning slides with abrasives such as rough tissues or other drying wipes. Slides with frosted ends are ideal for writing and labeling, particularly if slides are to be stored or added to a collection. Other considerations for slides are glass quality, thickness, and tendency to chip or break. Specialty items such as concave (an example is presented in Chapter 9), premarked, and prelabeled slides are available. For most work, simple inexpensive glass slides, 1 or 1.2 mm thick are adequate. For special applications, quartz slides (UV transparent), and plastic slides are available. Cover slips of several materials are available in a variety of shapes and sizes. Glass is the most commonly used material. The most important concern is usually thickness. The common numbering scheme is based on thickness:

No.	Thickness (mm)
1	0.13 to 0.17
1½	0.16 to 0.19
2	0.17 to 0.25

It is good practice to obtain or build your own set of tools for microscope work. To start, acquire a basic dissection kit that includes a probe or two, a razor blade or scalpel with replaceable blade, forceps, scissors, and a clear ruler. To supplement this, a tungsten needle probe is useful, along with a selection of microspatulas, brushes, microscribe (diamond tip), permanent and wax markers, pencils with good eraser ends, glass rods drawn to fine tips, and lens wiping paper.

5.2   MOUNTING MEDIA

When selecting a mounting medium, the goal is to optimize contrast and resulting visibility of the sample while collecting the most light possible at the objective lens. As noted in earlier chapters, optimal contrast is obtained when the refractive index of the sample is significantly different from the refractive index of the medium. However, recall that refraction occurs whenever light leaves one medium and enters another (such as air or glass). The refractive index of air is taken as 1.0, and that of glass in the slide and cover slip as about 1.52. Thus, any air–glass junctions in the light path will result in refraction (and dispersion) as shown in Figure 5.1. Selection of the mounting medium must be made with these principles in mind.

By keeping the refractive indices of the mounting media and the glass relatively close, refraction caused by the interface of sample and media (the interaction of interest) is delivered to the eyepiece with minimal distortion, as illustrated in Figure 5.2. Figure 5.3 shows the effects of mounting media on contrast.

For temporary mounting and sample examination, oils with known refractive indices are used. These oils are relatively viscous compared to water and allow easy particle manipulation with a tungsten needle. To prepare a temporary slide mount, place a small drop or two of the oil in the center of a slide using a rounded ball dropper or a drawn-out glass rod. Gently place or drop the sample into the oil. Carefully place a cover slip atop the sample and press down gently with the eraser end of a

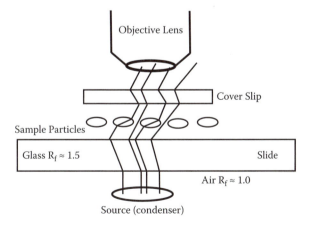

FIGURE 5.1   Light passing through a sample without mounting medium. Note that each junction results in refraction (greatly exaggerated here). The numerical aperture of the lens along with refraction determine how much light is collected by the objective.

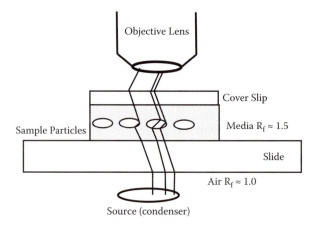

FIGURE 5.2   Same sample mounted in medium with RI similar to that of the glass in the slide and cover slip. The only two refractive junctions are (1) between air and glass and (2) between sample and medium.

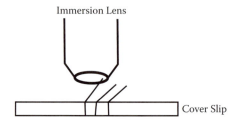

FIGURE 5.3   Effects of different mounting media on contrast. The glass standard used here has an RI of 1.52 and is mounted in a series of Cargille microscopy oils. Note that when the RIs of the sample and the oil are the same, the glass edges are all but invisible. The black circle shows the location of one piece of glass.

pencil. Any excess oil should be drawn off using an oil-absorbent cloth or towel. Press the cover slip down gently again to remove air bubbles.

A variety of products are available for creating permanent or semi-permanent mounts. Most work like paints or varnishes—by forming a polymer coating. Some media are water-based, others solvent-based, so it is important to determine whether a sample is compatible with the mounting material selected. For example, an aqueous-based medium will dissolve any water-soluble particulates. Conversely, if you are working with samples that can be dehydrated (e.g., biological samples), mounting media such as glycerin gels are required. If a medium is aqueous-based, pH compatibility must be considered, for reasons we will detail in Chapter 9. Ideally, you should check compatibility with a small representative portion of the sample before making a final medium selection. If the sample will be subjected to fluorescence studies, the background fluorescence of the medium must be considered as well. Many products are designed specifically for mounting specimens for fluorescent microscopy studies.

One of the oldest mounting media (still used) is a substance called Canadian balsam, a natural resin with excellent optical properties and a refractive index near that of glass. A later addition was Aroclor 5442 that belongs to a class of compounds called polychlorinated biphenyls (PCBs). Aroclor was a versatile medium, viscous, and with good optical properties. It has been phased out due to health and safety concerns.

Certain mounting media are considered semi-permanent. A commercial example is the Cargille Meltmount™ series, sold in small jars and as sticks. These products are waxy solids at room temperature and they liquefy under modest heat from a water bath or hotplate. After samples are placed in the liquefied melt, they re-solidify rapidly and are ready for observation. Should sample recovery be required, all that is needed is gentle heating, although it is difficult to completely remove the medium without resorting to a solvent rinse. Unlike the permanent mounts described above, melting media do not polymerize; they simply undergo a reversible phase change. Permanent mounts are usually created in polymer-based resins. These resins may cure via evaporative processes or by ultraviolet exposure. Many formulations include plasticizers to lessen stiffness. To prevent fading and cracking, antioxidants can be used.

Finally, a special type of oil for sample mounting called **immersion oil** is used specifically for and with high magnification immersion objectives. These objectives have high lens curvature and small numerical apertures (refer to Chapters 2 and 3). Immersion oils are formulated to carefully control refraction and allow maximum collection of light as shown in Figure 5.4. This is the reason that oil immersion lenses provide high magnification. The high curvature and lower numerical aperture resulting from high magnification can be partially offset by the use of immersion oil. The oil refracts light so that more of it is collected by the objective lens than would be collected if no oil was used. Some immersion objectives are designed to work with water rather than oil.

5.3 MICROMETERS AND MEASUREMENTS

To measure microscopic samples, the only requirements are a marked eyepiece (ocular) and calibration of the ocular. Because the calibration is magnification-dependent, each objective lens must be calibrated separately. A calibration is valid as long as the optical components of a microscope do not change. The tools needed

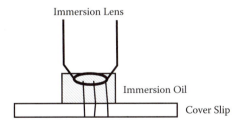

FIGURE 5.4   How an oil immersion lens works.

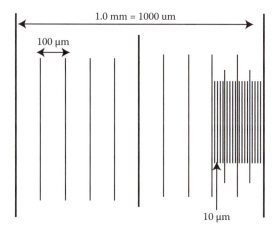

FIGURE 5.5   Stage micrometer.

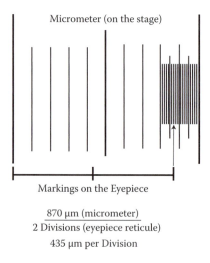

$$\frac{870 \ \mu\text{m (micrometer)}}{2 \ \text{Divisions (eyepiece reticule)}}$$
$$435 \ \mu\text{m per Division}$$

FIGURE 5.6   Calibration of eyepiece reticule using micrometer.

for the operation are a calibrated **micrometer** and an eyepiece reticule with calibration markings. The procedure is straightforward: the marks on the micrometer indicate a known distance that can be correlated with the markings on the eyepiece. The resulting calibration is expressed in units such as micrometers per division (a mark on the eyepiece).

Figure 5.5 shows a micrometer without graduation marks as seen through an eyepiece. Although there are different types of micrometers, most are marked similar to what is shown. The higher the magnification, the smaller the scale that must be used. Figure 5.6 shows a micrometer viewed through an eyepiece with division markings. Line up one of these reticule markings parallel to a major marking on the micrometer as shown. The distance per division on the eyepiece reticule is calculated based

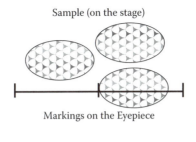

Sample (on the stage)

Markings on the Eyepiece

Width of Sample Particle ~1.7 Divisions
435 µm per Division
Width of Particle ~740 µm

FIGURE 5.7 Using calibrated eyepiece to obtain sample measurement.

on the known distance obtained from the micrometer. Of course, your microscope must be properly aligned and all objectives centered before beginning this operation. The procedure should be repeated for all your eyepiece–objective lens combinations. The only difference will be which scale is used; as magnification increases, you will need a smaller micrometer scale.

Use of calibration is illustrated in Figure 5.7. With the calibrated reticule installed, align a sample with the calibration marks and use the scaling factor determined to estimate the measurement of interest. Depending on how the eyepiece is marked, you may have to estimate readings, so report results cautiously using the appropriate number of significant figures. Also, it is good practice to measure many different particles from several perspectives to report an average over the collection rather than a single measurement. Finally, be sure to select the best magnification to perform the measurement. In the example provided, it might make more sense to repeat the measurement at the next highest magnification based on the amount of unmarked space on the eyepiece.

REFERENCES AND FURTHER READING

Y. Chen, Z.P. Guo, X.Y. Wang, and C.G. Qiu, Sample preparation, *Journal of Chromatography A* 1184: 191–219, 2008.

K.H. Esbensen, H.H. Friis-Petersen, L. Petersen, J.B. Holm-Nielsen, and P.P. Mortensen, Representative process sampling in practice: Variographic analysis and estimation of total sampling errors (TSE), *Chemometrics and Intelligent Laboratory Systems* 88: 41–59, 2007.

F.M. Garfield, Sampling in the analytical scheme, *Journal of the Association of Official Analytical Chemists* 72: 405–411, 1989.

P.M. Gy, Introduction to the theory of sampling 1: Heterogenity of a population of uncorrelated units, *Trends in Analytical Chemistry* 14: 67–76, 1995.

J.B. Holm-Nielsen, C.K. Dahl, and K.H. Esbensen, Representative sampling for process analytical characterization of heterogeneous bioslurry systems: A reference study of sampling issues in PAT, *Chemometrics and Intelligent Laboratory Systems* 83: 114–126, 2006.

M. Paakkunainen, S.P. Reinikainen, and P. Minkkinen, Estimation of the variance of sampling of process analytical and environmental emissions measurements, *Chemometrics and Intelligent Laboratory Systems* 88: 26–34, 2007.

L. Petersen and K.H. Esbensen, Representative process sampling for reliable data analysis: A tutorial, *Journal of Chemometrics* 19: 625–647, 2005.

L. Petersen, P. Minkkinen, and K.H. Esbensen, Representative sampling for reliable data analysis: Theory of sampling, *Chemometrics and Intelligent Laboratory Systems* 77: 261–277, 2005.

J.A. Springer and F.D. McClure, Statistical sampling approaches, *Journal of the Association of Official Analytical Chemists* 71: 246–250, 1998.

QUESTIONS

1. Assume you work in a forensic laboratory as a trace evidence analyst. Crime scene technicians deliver a bag from a vacuum cleaner used to collect trace evidence from a car. Describe the sampling challenge. How would you sample and how would you justify your approach?

2. Assume you have glass fragments on a glass slide, covered with a glass cover slip and mounted in a medium with the same refractive index (RI) as all of the glasses used. Draw a figure to show why you cannot see the glass sample.

3. Repeat the above exercise two more times. In the first case, assume the mounting medium has a lower refractive index than the sample. For the second, assume the RI of the medium is higher.

EXERCISES

Exercise 5.1: Particle size measurements

Materials needed:

Stage micrometer
Particulate samples such as prepared sets, starches (corn, rice), salt, sugar, etc.
Temporary mounting medium such as immersion oil or Cargille oil

Following the procedure outlined in this chapter, calibrate your microscope at all magnifications that you will use. Keep a record of the calibration as it remains valid until optical components are changed. Prepare a series of slides in temporary mounts (or dry mounts if you prefer) and experiment with obtaining measurements. One good exercise is to prepare a few slides of corn starch. Place a drop of oil on the slide and drop a few particles of the starch into the oil. Stir to spread out and apply a cover slip. Measure 10 to 20 grains (dimensions as the shape dictates) from each slide and determine the average diameter of a corn starch grain. Record the measurements on a spreadsheet and plot a histogram of the measurements. Note your observations. You may want to make permanent mounts as a reference set for practice and as a standard for future calibrations.

Exercise 5.2: Refractive index

Materials needed:

Set of ground glass and minerals of known RI (Cargille Laboratories' M1 set is ideal)

Set of oils of known RIs (Cargille Laboratories). You can substitute melting media as long as you have a variety of RIs to choose from (Cargille's RF-1/5 set is recommended; 0.01 intervals, 31 liquids, range of 1.400 to 1.700)

The Cargille set contains glass and minerals that cover a RI range from 1.34 to 2.40 in increments of 0.01 and is ideal for studying the effects of mounting media and refractive indices.

To start, select 10 glass and/or mineral samples that cover the range evenly. Select 10 refractive index oils that range from about 0.02 units below to about 0.02 units above the lowest and highest sample RIs.

Create a matrix in your notebook, recording the RIs of the samples along the top and the RIs of the mounting media on the bottom, then prepare slides using a tiny drop of oil, a few particles of sample, and a cover slip. Record what you observe for each combination in the matrix. Does a sample become invisible when certain combinations are used?

For a fresh set of experiments, select a glass sample with a refractive index around 1.56. Create a series of slides with oils beginning with RI values from 1.50 to 1.60 in units of 0.01. Notice as you focus on each sample where the line that defines the boundary between the particle and medium appears to move. This is called the Becke line, the movement of which can be used in RI measurements. For added practice, use your micrometry skills to measure the sizes of the particles. What dimensions make sense? Is it more useful to report an average size or a range? Sketch a few particles and show what dimensions you elected to record.

Exercise 5.3: Experiments with mounting media

Materials needed:

Variety of samples such as particulates, fibers, etc.
Swabs
Meltmount™ media 1.539 and 1.662
Organic-based resin such as Permount™
Ultraviolet curing resin such as Crystalmount™
Ultraviolet light (with proper protection from exposure)
Water-based media

Prepare several mounts of the same sample and cure as needed. This may take several hours, depending on the mounting media chosen. Be sure to label your slides completely (sample, medium name, medium RI, and preparation method). Keep track of what slides you have prepared. To obtain a biological sample, use a swab to collect cells from your inner cheek. Collate your observations and comment on the advantages and disadvantages of each medium.

6 Photomicrography

The recording of images, for the most part, is not handled by digital imaging techniques. The availability of cameras ranges from the adaptation of a relatively simple digital camera for recording images, to using a high quality camera system for analytical processing. In some cases an "image grabbing" card is required for the computer controlling the application. Acquisition of the image by using filters and the operation of a charge-coupled device (**CCD**) or chip play important roles in image acquisition. The subsequent processing of images is performed using software; packages such as Adobe Photoshop®,* ImageJ,† and GIMP‡ are readily available. The latter two packages are available under a general public license for free downloading.

This chapter focuses on basic image acquisition and processing. Image acquisition will be demonstrated using a digital single lens reflex (**SLR**) camera. Basic image processing techniques for enhancement and analysis will be covered. Remember, good images require good illumination, so be sure to align your microscope as per Chapter 4 before acquiring images.

6.1 ACQUISITION

The general requirement for set up is a **trinocular head** attached to a recording device (it is sometimes possible to attach a camera to an eyepiece if a trinocular head is not available). An intermediate lens, camera (software control if required), and adaptors are needed for capturing images. It is possible to make use of webcams, but they typically do not provide images of sufficient quality.

Imaging through a microscope has come a long way since people first pressed camera lenses to eyepieces. In today's digital world of "disposable" images, it is common to capture images of everything you see, then keep the best, and discard the rest. The process was more complicated when film-based cameras were used in pre-digital times; imaging then was as much art as science. It should be obvious, however, that the relative ease of imaging for documentation purposes does not suggest that sloppy or poor images are acceptable. Imaging for quantitative or analytical purposes can be quite different. The tools should suit the application. This chapter will show you relatively simple ways of capturing images using common digital SLR cameras. Today's ever changing digital camera world, where new models appear frequently, still requires knowledge of basic photographic techniques.

We will discuss a camera relative to a microscope, since both are optical imaging devices. Recording an image through a microscope is known as **photomicrography** (and not micro-photography). A digital SLR camera consists of a few basic

---

* http://www.adobe.com/products/photoshop/compare/
† http://rsbweb.nih.gov/ij/
‡ http://www.gimp.org/

components: objective lens (with built-in aperture), shutter curtain, and imaging sensor, along with a flip-up mirror* and a **pentaprism** to view imaging through the lens (TTL). It also has a light sensor to measure the amount of light reaching the sensor and a microprocessor to control the camera. All in all, the device is very complex. The correct exposure of an image is controlled by the speed of the shutter and the size of the aperture. In normal photography the aperture can be used to control the depth of field of an image. When a camera is attached to a microscope, the lens is removed and a tube is attached to the trinocular head. An intermediate lens is inserted to ensure that the image is focused on the sensor of the camera. Some microscopes require adjustment of the length of the imaging tube to ensure crisp images.

Some of the more complex and expensive digital SLR devices allow direct imaging thorough computer control (e.g., the Nikon D300 and its Camera Control Pro software). The advantages of these systems include the ability to view an image directly before collection, the use of a relatively high resolution sensor, and the capability of adjusting white balance and color control before the image is acquired.

6.2   IMAGE CAPTURE

If you prepare images regularly, it is advisable to purchase acquisition software for your camera. Appropriate software is available from most major digital SLR manufacturers (e.g., Nikon and Canon). The software will allow you to control your camera through your computer (Figure 6.1). The camera simply connects to your computer through a USB or **Firewire** cable. Since the camera is connected to the microscope without a lens, you are obviously unable to control the f-stop of the camera since most apertures are integrated into lenses. Setting your camera to manual mode will allow exposure control via camera shutter speed. Nikon's software package allows compensation for exposure.† This allows you to take a series of images and vary the shutter speed to obtain the desired outcome.

These types of cameras and the relevant software will enable you to produce a live image on your computer (Figure 6.2) that shows the field of view of the camera (as opposed to the view from an eyepiece). You can also "tweak" the focus (especially with high NA objectives) to achieve the focal plane you desire.

6.3   BASIC IMAGE PROCESSING

We will provide some basic principles of image processing and use (with thanks to Photoshop, GIMP, and ImageJ). Whatever system you use, it is important that you understand the effect of the processing to be undertaken. The algebra related

---

* Most digital SLR cameras will allow for viewing the scene with the aperture completely open. Some models will allow for a preview with the aperture stopped down to that which will be used in the photograph. In order to make use of the TTL feature, a mirror is needed to reflect the image from the lens up into the pentaprism. When the release is pressed on the camera, the mirror flips up to allow the light to reach the sensor and the aperture stops down to its correct setting. The shutter curtain opens and closes to ensure the image is correctly exposed.

† http://www.nikonusa.com/Find-Your-Nikon/Product/Imaging-Software/25366/Camera-Control-Pro-2.html

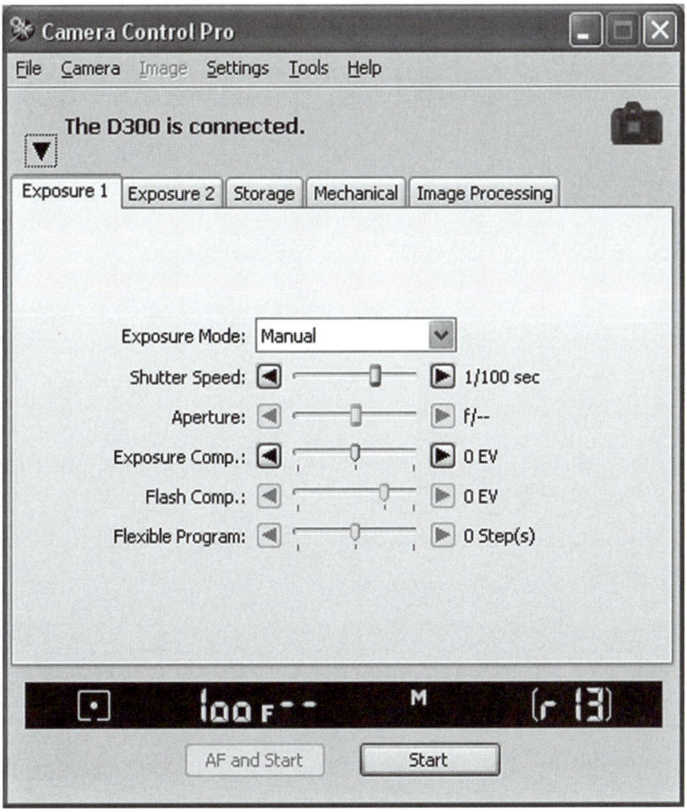

FIGURE 6.1 Dialog of camera control software. The shutter speed, exposure compensation, and capture type can be controlled by the PC.

to image processing can be complex and it can be easy to lose your way. The primary rule is that image processing should not be used to enhance an image that was poor in the first place. The best steps to achieve better images are (1) efficient and correct sample preparation and (2) a thorough understanding of your optical instrument.

Images can be collected in many different formats. Most basic cameras will record images in a **JPEG** (Joint Photographic Experts Group*) format representing a class of images that use a **lossy compression algorithm**. This means that the image is re-sampled to reduce its size. Adjacent pixels may be of the same or very similar color, so the algorithm will give that area a single color, thus reducing the amount of information it has to store. This may be acceptable for party shots, but is not suitable for documentation.

If your camera has only the ability to record JPEGs, they should be converted, at a minimum, to lossless file types such as TIFF (tagged image file format) as a first step in processing. With both image types, the camera will pre-process the image before

---

* http://www.jpeg.org/

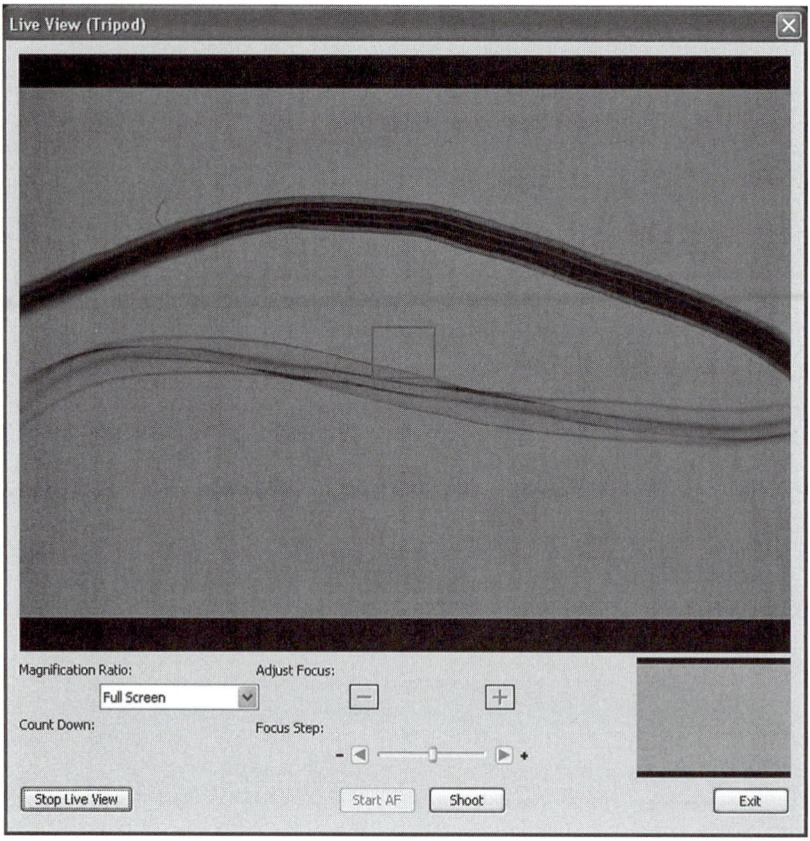

FIGURE 6.2 Some software allows a live view of an image. The advantage of this feature is that fine adjustments can be made to the focus setting. If this feature is unavailable, a number of images can be taken and a cursory evaluation for quality can be performed with software such as Picasa (http://picasa.google.com/).

saving it to a media card. With a digital SLR camera, you will usually be able to save the images in a **raw format**. This is the image taken by the camera without pre-processing. Adobe Photoshop's Bridge®* can convert raw images into a usable format. A raw image can serve as a type of negative (as in film cameras). The problem with the raw format is that it is proprietary to each camera manufacturer, so a user must have the software to convert the image so that it can be opened by an image processing program. Adobe introduced a new (publicly available) format known as the Digital Negative (DNG).[†] It is anticipated that more camera manufacturers will adopt this format to allow for an open standard for the raw files created by different cameras. Another advantage of the DNG format will be that the images will be available for use in the future.

---

* http://www.adobe.com/products/creativesuite/bridge/bridgehome/
[†] http://www.adobe.com/products/dng/

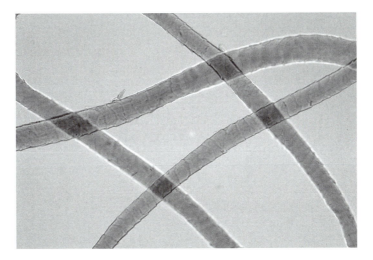

FIGURE 6.3 Photomicrograph of woolen fibers under bright field conditions.

A few concepts should be clarified before we continue to image processing. The first is image size. The histogram for Figure 6.3 (which is shown in Figure 6.4) indicates a distribution of all of the **pixels** in an image based on their grayscale of black to white (0 to 255).This histogram represents a combination of all colors in the image shown in Figure 6.3. A histogram tells us how many pixels in an image have a specific grayscale. It does not indicate where these pixels are located in the image. The most common grayscale images exhibit 256 shades of gray.

To provide a better understanding of image size, we must look at a binary measure of grayscale (and color) images. We can rewrite 256 as 2^8 ($2 * 2 * 2 * 2 * 2 * 2 * 2 * 2$). The exponent indicates an 8-bit image. A 16-bit image will therefore have 2^{16} shades of gray or 256 times as many shades as an 8-bit image ($2^{16} = 2^8 * 2^8 = 256 * 256 = 65,536$). Another way of conceptualizing this arrangement is that for each of the 256 grayscales in an 8-bit image there are another 256 grayscales in a 16-bit image. We can consider a color (RGB) image as a combination of three images (one each for red, green, and blue) and thus a 24-bit image ($3 * 2^8$ or 2^{24}).

The most common initial processing step is the levels function applied to an image to increase the contrast within the image. A histogram may not be evenly spread out over the 256 grayscales. The image is adjusted so that all of the grayscales are used (see Figure 6.5). This may appear to improve the image, but does so at a cost. Compare the before and after histograms of an image whose levels have been adjusted.

Notice the gaps in the distribution. Gaps arise because the distribution is limited to the 256 levels and is not continuous. In this particular case, levels 163 to 255 have been redistributed over the 0 to 255 scale, leaving 62 levels that have no pixels. The second step is to adjust the brightness. Actually both contrast and brightness change in the levels dialog.

Figure 6.6 is the brightness and contrast dialog in ImageJ. The histogram once again is Figure 6.4. To set the brightness and contrast, you can use the minimum and maximum or the brightness and contrast sliders. The contrast slide will change the

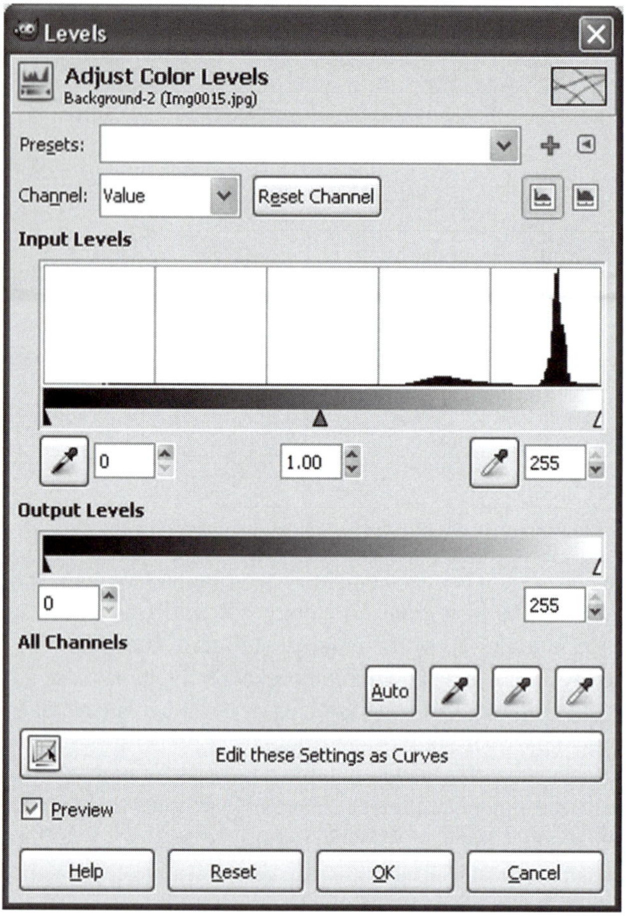

FIGURE 6.4 Dialog to adjust color levels in GIMP. This dialog displays the histogram of the image in Figure 6.3.

gradient of the line in the figure and the brightness slider will adjust the X intercept. This line will map the 0 to 255 grayscale pixels in the input image to a new set of grayscale images as determined by the line.

The image can also be corrected for color, but for photomicrographs this is better done during acquisition. The color correction function in Photoshop will remove casts. This is done by selecting black, white, and gray sections of the image using the dropper tool (Figure 6.7).

Further improvements may be effected by sharpening or removing noise. Noise can be removed by a linear filter. We can think of our image as a matrix of size x,y. We can apply a linear filter that can be thought of as a matrix that is typically 3×3 or 5×5 in size. The algebra for processing is slightly different from normal matrix algebra. In this case, we consider the central element of our operating matrix as the focal point. We place the matrix slightly in from the edge of the image so that a pixel

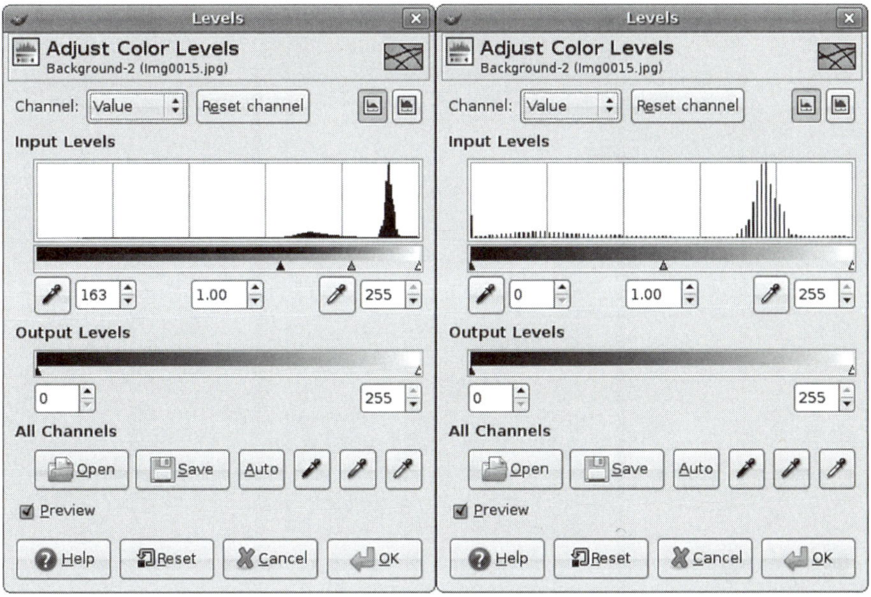

FIGURE 6.5 Adjust Color Levels dialog in GIMP with new input levels selected (left), and the same dialog after application of these settings (right). Note how the histogram appearance has changed.

value of the image is related to the values in all cells of our operating matrix. The resultant value we calculate will become the new value for the central element of the image:

$$\begin{bmatrix} 1 & 1 & 1 \\ 1 & 1 & 1 \\ 1 & 1 & 1 \end{bmatrix} \begin{bmatrix} 24 & 28 & 37 \\ 17 & 36 & 28 \\ 13 & 16 & 22 \end{bmatrix} = \begin{bmatrix} 24 & 28 & 37 \\ 17 & 36 & 28 \\ 13 & 16 & 22 \end{bmatrix}$$

$$= 1*24 + 1*28 + 1*37 + 1*17 + 1*36 + 1*28 + 1*13 + 1*16 + 1*$$

$$= 221$$

We obviously would like to have a new 8-bit image, so we normalize the data by dividing by the sum of the elements of our operating matrix (9) to yield the result (24.5 which becomes 24).

Image analysis focuses on making measurements. This can be preceded by some processing to identify edges. An edge can be defined as a relatively rapid change in grayscale in a particular direction. Edges can also be detected with a filter. Many types of filters may be applied, but for the sake of simplicity we focus on one type to describe the concept. A gradient filter will calculate the average gradient in an image across the operating matrix. We will consider a filter known as a **Prewitt**

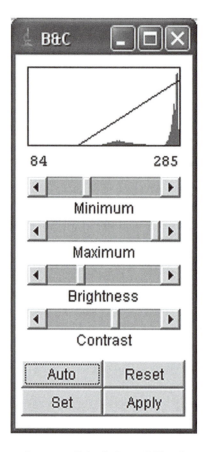

FIGURE 6.6 Brightness and contrast dialog in ImageJ. The slope and intercept of the transformation line can be adjusted in a number of ways to achieve optimal results.

operator—a 3×3 matrix. This size was chosen to eliminate the possibility of noise interference from evaluating a single line:

$$\begin{bmatrix} -1 & 0 & 1 \\ -1 & 0 & 1 \\ -1 & 0 & 1 \end{bmatrix} \text{ for horizontal direction}$$

$$\begin{bmatrix} -1 & -1 & -1 \\ 0 & 0 & 0 \\ 1 & 1 & 1 \end{bmatrix} \text{ for vertical direction}$$

This filter actually performs a smoothing operation in one direction before calculating the gradient. Our image of the fibers was processed by both filters to create two resultant images. The two images were added together, and the contrast and

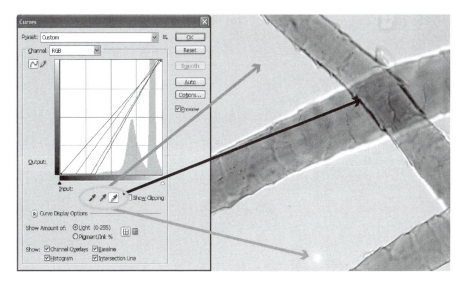

FIGURE 6.7 Correcting color using the curves dialog in Photoshop. The white, gray, and black droppers are used to select the corresponding colors within the image to correct the color of the image.

FIGURE 6.8 Vertical and horizontal Prewitt filters were applied separately to two iterations of Figure 6.3. The resultant images were added together, the contrast and brightness of the summed image was adjusted, and finally the image was smoothed.

brightness adjusted and smoothed to remove noise. Figure 6.8 depicts the final smoothed image.

As discussed in Chapter 5 (Measurements and Micrometry), the use of measurements to determine properties such as birefringence (Chapter 7) is an important requirement. If the size of a particular feature in an image can be determined, the data can be used to calibrate the image and make multiple measurements easy. If

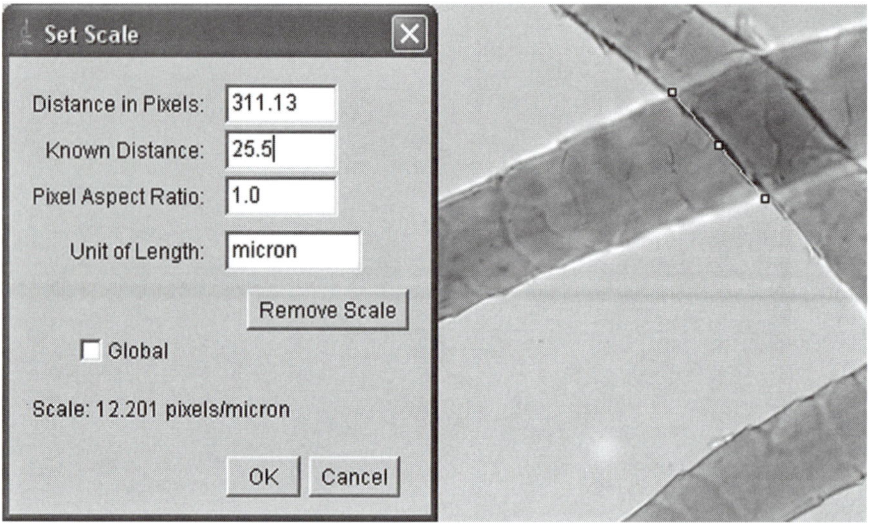

FIGURE 6.9 Any scale of an image can be set in ImageJ. The actual dimension of the object is related to the number of pixels. Actual measurements (length, area, angles, etc.) can then be made.

the specific thickness of one fiber in our image is known we can calibrate the image using the *Set Scale* function in ImageJ (Figure 6.9). After the scale is set, multiple features (width, length, angles, diameter, area, etc.) can be measured directly via the region of interest (**ROI**) manager. See Figure 6.10. Other features contained in these software packages can provide for automated measurements. The best approach is to download and evaluate the software by experimentation.

When imaging with high NA objectives, depth of focus is limited. This can be eliminated to some extent by recording a few images at different levels through the sample and then adding the images to form a composite image. This can be seen in Figure 6.11. A sample of **trilobal** nylon carpet fibers was mounted for examination. In the first image, the focal point was the top of the upright lobe. The second image was focused at the center of the fiber and the final image was focused at the base of the fiber. The three images were added together to produce the final image using the *Process > Image Calculator* functionality in ImageJ.*

---

* Remember to check the 32-bit (float) result box. After completion, convert your image back to an 8-bit grayscale image.

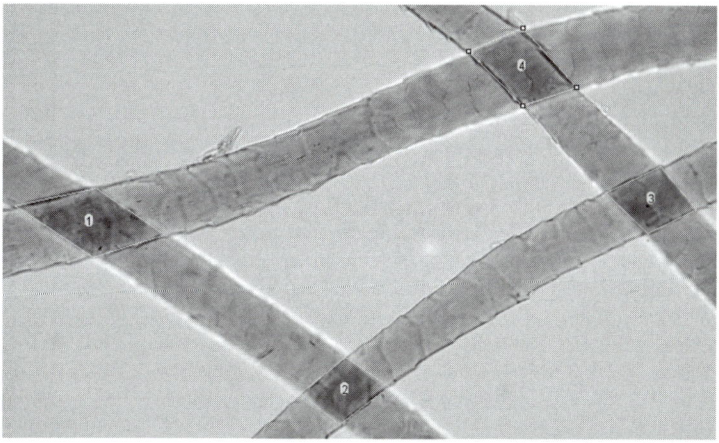

Label	Area (microns^2)	Perim. (microns)	Grayscale Analysis			
			Mean	StdDev	Min	Max
1	397.2	82.63	135	24	49	231
2	312.7	71.38	131	21	24	224
3	243.9	63.08	126	22	61	241
4	1090	159	177	22	95	255

FIGURE 6.10   Image with four regions of interest (ROIs) indicated. A number of measurements (in micrometers) have been determined by ImageJ.

FIGURE 6.11   Performing image addition to improve the depth of field of the final image.

REFERENCES AND FURTHER READING

W. Burger and M. Burger, *Digital image processing: An algorithmic introduction using Java*, Springer, 2008.

J. Delly, *Photography through the microscope*, Eastman Kodak, 1988.

R. Ditchburn, *Light*, Blackie & Son, 1963.

M. Felsberg, in *On second order operators and quadratic operators*, IEEE, 2008.

R. Fischer, T.G.B. Yoder, and P. Yoder, *Optical system design*, SPIE Press, 2008.

R. Gonzalez and R. Woods, *Digital image processing*, Prentice Hall, 2007.

D. Halliday and R. Resnick, *Fundamentals of physics*, John Wiley & Sons, 1981.

R. Kirsch, Computers determination of the constituent structure of biological images, *Computers and Biomedical Research,* 1971, 4: 315–328.

D. Murphy, *Fundamentals of light microscopy and electronic imaging*, Wiley-Liss, 2001.

A. Peck, *Beginning GIMP*, Apress, 2006.

G. Reis, *Photoshop CS3 for forensics professionals*, Sybex, 2007.

J. Robertson, *Forensic examination of hair*, Taylor & Francis, 1999.

J. Russ, *Image processing handbook*, CRC Press, 2007.

QUESTIONS

1. Why is it difficult to focus an image through a microscope when a high numerical aperture objective is used?
2. If your camera records only JPG images, why is it important to convert them to a TIFF format as a first step?
3. Why would you prefer to record an image as 16-bit, even if image processing software can work only with 8-bit images?
4. The color balance in an image can be adjusted by the camera. In photomicrography, in what other way can it be adjusted?
5. We know that the median of a group of numbers is the middle number when the group is sorted in ascending order. Explain how you think a 3 × median filter would work.

EXERCISES

Exercise 6.1: Imaging large specimen using high NA objective (similar to Figure 6.1).

Materials needed:

> Microscope set up under Kohler Illumination (with 60× objective or similar)
> Sample of a carpet fiber with a multilobal cross-section
> Camera set up on microscope
> ImageJ software

Download and install the ImageJ software. Set up Kohler Illumination on your microscope with the 60× objective. After you have achieved the correct balance, take three images of your fiber. The three images should represent three optical sections of the fiber, in other words, three separate sections of the fiber in focus in each image.

These three images must be added to each other to create a final image. Open each image and then add the first two by using *Process > Image Calculator.* Select one as *Image1* and another as *Image2.* Select the *Add* operation, and ensure that the *Create New Window* and *32-bit (float) Result* boxes are checked. The *32-bit (float) Result* box is checked to prevent clipping the image (if two pixels each had a value of 186, then their sum is 372; this is greater than 255, so the resultant value remains 255). We will convert the result to a grayscale image when completed. Add the resultant image to the third original image. The final image can be smoothed, and the brightness and contrast adjusted.

Compare your final image to the three original images. Evaluate the DOF and compare it to the DOF you can calculate using Equation 2.11 (to do so you must insert a band pass filter to know the wavelength of the light used for illumination). This can also be estimated by using the green channel of the images by splitting the channels of each of the initial images using the *Image > Color > Split Channels* function and then selecting only the green channels.

Exercise 6.2: Applying filters to images using Adobe Photoshop

Materials needed:

Images of objects with well-defined and ill-defined edges
Adobe Photoshop (ImageJ can be used but requires programming)

Open image using Adobe Photoshop. Both Photoshop and ImageJ include a number of predefined filters that can be applied directly. If you wish to experiment, open the *Filter > Other > Custom* dialog. You will be presented with a 5 × matrix whose central element will represent the pixel of interest. A black cell in the filter represents a value of zero. Enter the values for each cell. Note that scale value is a normalizing factor. It should equal the sum of all the cells you enter (ignore the offset box). The following are Kirsch (edge) filters:

$$\begin{bmatrix} -1 & 0 & 1 \\ -2 & 0 & 2 \\ -1 & 0 & 1 \end{bmatrix}, \begin{bmatrix} -2 & -1 & 0 \\ -1 & 0 & 1 \\ 0 & 1 & 2 \end{bmatrix}, \begin{bmatrix} -1 & -2 & -1 \\ 0 & 0 & 0 \\ 1 & 2 & 1 \end{bmatrix}, \ldots$$

Notice how the filter rotates 45° with each step. You can calculate the next five filters to complete the set. ImageJ allows you to add all the filtered images together to form a final image that emphasizes all the edges. Other filters that can be evaluated are

Roberts (edge):

$$\begin{bmatrix} 0 & 0 & 0 \\ 0 & 1 & 0 \\ -1 & 0 & 0 \end{bmatrix}, \begin{bmatrix} -1 & 0 & 0 \\ 0 & 1 & 0 \\ 0 & 0 & 0 \end{bmatrix}$$

Scharr (edge):

$$\begin{bmatrix} 3 & 10 & 3 \\ 0 & 0 & 0 \\ -3 & -10 & -3 \end{bmatrix}, \begin{bmatrix} 3 & & -3 \\ 10 & & -10 \\ 3 & & -3 \end{bmatrix}$$

Laplacian operator (sharpening):

$$\begin{bmatrix} 0 & 1 & 0 \\ 1 & -4 & 1 \\ 0 & 1 & 0 \end{bmatrix}, \begin{bmatrix} 1 & 1 & 1 \\ 1 & -8 & 1 \\ 1 & 1 & 1 \end{bmatrix}, \begin{bmatrix} 1 & 2 & 1 \\ 2 & -12 & 2 \\ 1 & 2 & 1 \end{bmatrix}$$

Smoothing:

$$\frac{1}{9}\begin{bmatrix} 1 & 1 & 1 \\ 1 & 1 & 1 \\ 1 & 1 & 1 \end{bmatrix}$$

7 Polarized Light Microscopy

One of the most powerful microscopy techniques is also one of the simplest. Polarizing light microscopy (PLM), also called petrographic microscopy, was first used in geology for identification of minerals. Samples that interact with polarized light are crystalline or pseudocrystalline, meaning that some order appears within their structures. The degree of order is revealed by interference colors created by interactions of light with the material. Light may be polarized in several ways: linearly, circularly, and elliptically. Our discussion will focus on linear polarization. It is a good idea to return to Chapter 1 to review the concepts of vector and wave notation as we will make extensive use of these concepts in this chapter.

All the principles discussed up to this point regarding optics, illumination, Kohler Illumination, and microscope set up apply to PLM just as they do to bright field microscopy. As we will see, turning a bright field microscope into a PLM microscope is a simple proposition. Therefore, be sure that your microscope is properly aligned and optimized for PLM work. This is particularly important for photomicrography because PLM produces bright vibrant interference colors that are used diagnostically.

7.1 PLANE POLARIZED LIGHT

Visible light (electromagnetic energy) emitted from a source such as the sun or a lamp vibrates in all directions, as illustrated in Figure 7.1. When light is treated as an oscillating wave of electromagnetic energy, the **direction of propagation** describes the direction the wave front is moving in space. The **direction of vibration** describes how the wave is oscillating (here, up and down in the plane of the paper). When the light is unpolarized, the direction of propagation is constant, but only one plane of vibration is detected.

Most light sources used in microscopy are unpolarized. Recall from Chapter 1 (Section 1.1) our discussion of how and why a tungsten filament emits light. From any one electron, the emission is polarized in the direction perpendicular to the plane of motion. However, since all the valence electrons are emitting and their orientations are random, the intrinsic polarization of the emission of any single electron is lost. Over the collection of emitting electrons, the light is unpolarized and vibrating in all planes. There are several methods of depicting this, as shown in Figure 7.1. In some cases, it is convenient to look at the waveforms, and in others, the vectors, as will be shown in the following sections.

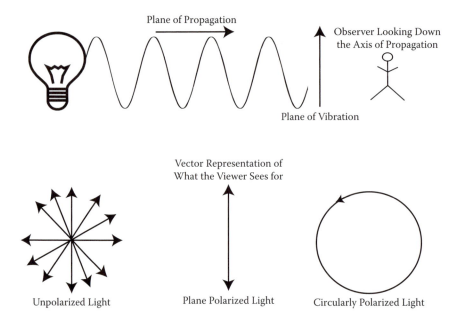

FIGURE 7.1 Polarized light as it would appear to an observer. Refer to Chapter 1 for a review of vector notation.

To polarize light from a non-polarized source, all that is needed is a filtering device that can remove all planes of vibration save one. If a filter is placed in the light path such that only one direction of vibration is allowed, the light is said to be polarized. Using a polarizing filter does not change the constituent colors of the light, nor does it alter any of the wavelengths, frequencies, or speed of propagation. However, the intensity of the light is decreased since some of the energy is filtered out. The effect is demonstrated by polarized sunglasses. When bright light from the sun reflects off of a surface such as a window, glare results. Polarized sunglasses reduce this glare by blocking all vibrational directions except one. When viewed through polarized lenses such as in sunglasses, the appearance of the window is not changed, nor is the color. The polarizing filter creates light that is said to be "plane polarized" because the electromagnetic radiation oscillates in only one plane. Similarly, a polarizing filter can be mounted on a standard light microscope such that only polarized light is directed through the sample. The viewer may notice that the field of view is slightly less bright because less total light reaches the eyepiece. Commercial polarizing devices are of several types including wire polarizers, films, and crystal-based systems. Crystal-based polarizers merit discussion because they are useful for explaining how polarized light can be used diagnostically to analyze crystalline and pseudocrystalline structures.

One of the earliest materials used to construct polarizers was the calcite ($CaCO_3$) mineral. Calcite's crystal falls into the hexagonal scalerohedral or rhomboidihedral class and may be compared informally to a faceted saucer shape. Compare this to Figure 7.2, the crystal structure of sodium chloride (NaCl). All faces of

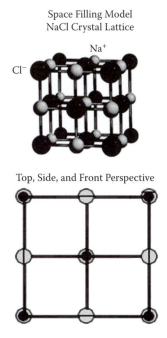

FIGURE 7.2 Crystal structure of sodium chloride (NaCl). The arrangement appears the same no matter which axis the viewer looks down. (Images adapted from the Crystal Lattice Structures Web page, http://cst-www.nrl.navy.mil/lattice/, provided by the Center for Computational Materials Science of the United States Naval Research Laboratory.)

NaCl appear the same and, as a consequence, light entering the crystal experiences the same atomic environment regardless of which direction it travels.

This is not the case with calcite (Figure 7.3). The view from the top is very different from the frontal view, but that organization is repeated throughout the crystal. Consider how this crystal would interact with light propagating through it from different directions. If light impinged on the front surface and passed through, the electromagnetic environment that the oscillating wave encounters would be different from the environment that a wave of light traveling from top to bottom would encounter. Because the crystal structure is ordered and repetitive, the relative differences remain constant. As a result, the refraction of light traveling from top to bottom is different from the refraction of light traveling front to back.

A material such as calcite is called **birefringent** because it has two different refractive indices that depend on the spatial orientation of the crystal relative to the light source. Put another way, calcite is an example of an **anisotropic** material (meaning literally *not the same*). Sodium chloride is an example of an **isotropic** material—it has only one index of refraction. Glass falls into this same category but for a different reason. Glass lacks an organized crystal structure; it has no directionality and interactions with light are random in all directions. Anisotropic media have two or more refractive indices. By definition, a birefringent material is

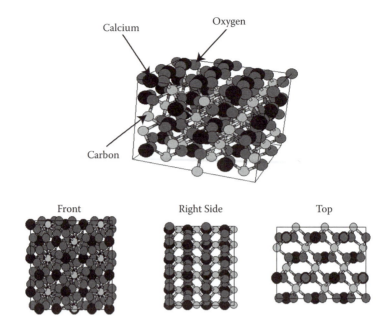

FIGURE 7.3 Crystal structure of calcite ($CaCO_3$). The arrangement is not the same on all axes. (Images adapted from the Crystal Lattice Structures Web page, http://cst-www.nrl.navy. mil/lattice/, provided by the Center for Computational Materials Science of the United States Naval Research Laboratory.)

anisotropic, but the reverse is not necessarily true; a material may have more than two refractive indices.

Every anisotropic crystal exhibits one axis or direction along which light will propagate with the same velocity regardless of what plane the light is polarized in and this is called the **optic axis.*** This axis may have nothing to do with the shape or dimension of the crystal; for example, a rectangular crystal may have an optic axis that runs diagonally. We can always find the direction of the optic axis based on how light interacts with it. When discussing how polarized light interacts with a crystalline material, the optic axis serves as an important reference. If plane polarized light propagates through a crystal parallel to the optic axis, the material appears isotropic; no beam splitting occurs. If light propagates down any other axis, the beam will be split.

Before discussing the details of polarized light interactions with materials, some notation conventions must be addressed. The complexity of the interactions we study in polarized light microscopy is often difficult to illustrate on paper using wave notation. As shown in Figure 7.4, waves of electromagnetic energy can be depicted in two dimensions using symbols as shown. In the middle diagram, a wave of light vibrating in the plane of the paper is propagating from left to right.

---

* Some crystals are biaxial and have more than one optic axis, but we will focus on single or uniaxial crystals here. The next chapter delves into more detail on crystallography. For more information, consult the references listed at the end of the chapter.

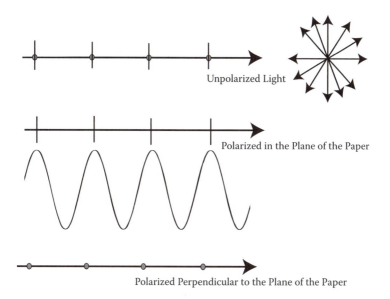

FIGURE 7.4 Notation conventions for polarized light.

This beam is polarized since there is only one plane of vibration. The short lines correspond to wave crests and the arrow indicates the direction of propagation. The lower diagram depicts a polarized beam vibrating in and out of the paper perpendicular to it. The diagram at the top shows how unpolarized light (Figure 7.1 also) is shown using this scheme.

7.2 DOUBLE REFRACTION AND CALCITE

To understand how polarized light interacts with calcite, imagine that you are deep inside a crystal and observing a beam of polarized light interacting with a point in the crystal lattice. As long as you are not looking down the optical axis, you would observe the beam splitting into two mutually perpendicular components or wavefronts (Figure 7.5). Recall that the degree of refraction in two media depends on the difference in velocity. Here, the light is not moving in two different media, but it is propagating in two directions within a crystal. Along the two paths, the light encounters two different electromagnetic environments and the velocity in one direction is slower than in the other. Viewing the spot head-on where the divergence occurs (the star in Figure 7.5), you would see one wave move faster than the other. These can be identified by the speed difference as **fast** and **slow** waves (or components). These terms are used in polarizing light microscopy.

Defining the other terms requires an understanding of how the beam splits. Calcite again provides a good starting point. Under ordinary illumination (unpolarized light), clear calcite displays a phenomenon called **double refraction**, resulting from the beam splitting behavior of the crystal. As seen in Figure 7.6, unpolarized light passes from the sample and enters the calcite crystal at an angle oblique to the

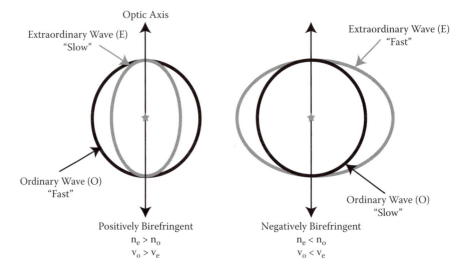

FIGURE 7.5   Point source of light within a birefringent crystal. The star indicates the interaction (described in text). Two wavefronts propagate from this point source. The speed of the propagation depends on the refractive index experienced by each wavefront.

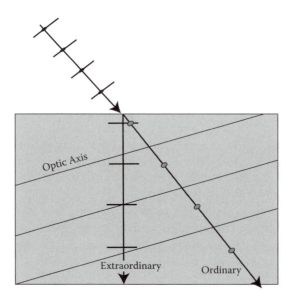

FIGURE 7.6   Beam splitting in calcite (unpolarized light). Note that the beam is split into two components that exit the crystal at different locations.

optic axis. The beam splits and the two rays vibrate perpendicularly to each other. When these two rays emerge and are seen by the viewer, two images are evident, one created by the **ordinary ray** (O) and one by the **extraordinary ray** (E). The trajectory of the O ray is named so because it follows the usual (ordinary) rules of refraction. The same is not true of the E ray that veers off in a different direction because of the different refractive index it experiences.

As a result of the split and interaction with the crystal, the O and E rays are linearly polarized and vibrate in mutually perpendicular directions. If you were poised over a sample and viewed it through a calcite crystal, you would see two images, one created by the O ray and the other by the E ray. Interestingly, if you were to rotate the crystal, the image created by the O ray would appear stationary while the image created by the E ray would appear to circle around the other. This makes sense when you remember that the ordinary ray will always be refracted normally and the E ray will change as a function of the angle to the optic axis.

7.3 PLANE POLARIZED LIGHT AND BIREFRINGENT MATERIALS

When plane polarized light interacts with a birefringent material such as calcite (Figure 7.6), the crystal still acts as a beam splitter as described above. The beam splits into two components that vibrate perpendicularly to each other. As noted above, if the axis of propagation light is parallel to the optical axis, the crystal appears isotropic. At the other extreme, if the angle of incidence is perpendicular to the optic axis, the beam splits, but both propagate at the same speed and emerge at the same time. When recombined, the appearance of the sample is not altered even though the beam has split.

These phenomena become interesting (and useful) under polarized illumination when the angle of incidence is oblique relative to the optic axis of the sample, as was the case with the calcite shown in Figure 7.6. When the beam splits into two rays that vibrate perpendicularly to each other, each ray experiences a different electromagnetic environment and propagates at a different speed. As a result, one component falls behind the other, a phenomenon called **retardation**. In the situation illustrated in Figure 7.6, the beams emerge at different places and thus produce two different images. However, suppose that instead of being viewed as two separate beams, the two beams were recombined. As we discussed in Chapter 1, interference occurs whenever light interacts with itself. The interference pattern produced by birefringent materials is a core concept in polarizing light microscopy.

When the emerging beams are recombined, interference occurs because the two waves are out of phase relative to each other. The viewer sees interference colors that relate to the type of material and its thickness. The thicker the sample, the greater the retardation of the slower beam relative to the faster beam. We will later see how we can take advantage of this effect to make interference colors more vivid and diagnostically useful. It is important to remember that the colors seen under these conditions do not represent the intrinsic color of the sample. They are artificial colors created by the interaction of the light with the sample. Interference colors are seen frequently outside the laboratory, such as on soap bubbles and thin oil coatings on water.

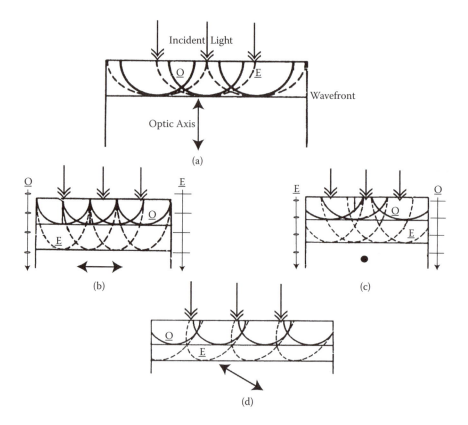

FIGURE 7.7 Different ways to denote the interaction of light with calcite as a function of angle of incidence relative to the optic axis.

Figure 7.7 reiterates this concept using calcite, a negatively birefringent material, as the example. Note that in the top frame (a), the incident light is propagating parallel to the optic axis and no retardation of the O or E rays occurs. In (b), the incident light is propagating perpendicular to the optic axis, the O component vibrates in the plane perpendicular to the paper, and the E ray vibrates in the plane of the paper. Since the birefringence is negative, the E ray becomes the faster component. In (c), the opposite occurs when the optic axis is sticking up from the plane of the paper. Finally, in (d), the angle of incidence is oblique. For PLM purposes, this is important because we rotate and move samples on a microscope stage and thus manipulate the angle of incidence of the polarized light relative to the optic axis of the crystal.

It is important to emphasize that double refraction and birefringence are related but are not the same things. Both relate to the ability of a crystal, via its ordered chemical structure, to act as a beam splitter. However, double refraction is what we see looking through calcite—two rays and two images. Birefringence is what causes double refraction; a variation in the refractive index that is a function of direction in a crystal. This too arises from the order of the crystal structure.

Birefringence is useful in microscopy because it can be measured. The numerical difference in the two refractive indices (O and E) is called birefringence (B or Bi) and is calculated as

$$B = n_e - n_o \qquad (7\text{-}1)$$

Materials with low birefringence show small differences, whereas highly birefringent materials exhibit large numerical differences. Materials (e.g., calcite) can be negatively birefringent or positively birefringent, as shown by the accompanying difference in sign. The parallel and perpendicular notations are also seen in this context and the equation is

$$B = n_{\updownarrow} - n \qquad (7\text{-}2)$$

This notation refers to the way the polarizing elements in the microscope are arranged, a topic to be discussed in the next section. Figure 7.5 illustrates the difference between positively and negatively birefringent samples as related to the O and E components.

It is important to remember that the light discussed here that illuminates a sample or passes through a crystal consists of many wavelengths, even though the figures infer a single waveform for the sake of simplicity. In polarized light microscopy, we work with all wavelengths of light in the visible spectrum and as a result, interference patterns will become more complex. Because refractive index depends on wavelength, so too does retardation. As an example, Table 7.1 summarizes refractive indices for calcite as a function of wavelength. This wavelength-dependent difference plays a central role in generating interference colors.

7.4 POLARIZING LIGHT MICROSCOPES

Polarizing light microscopes represent variations of standard compound microscopes. The primary difference is the presence of two polarizing filters. The first, called the **polarizer**, is placed in the light path between the lamp and the sample. The second, called the **analyzer**, is placed after the sample and before the eyepiece. Either or both filters may be rotatable.

TABLE 7.1
Refractive Indices of Calcite

Wavelength	n_o	n_e
242	1.7811	1.5378
410	1.6801	1.4964
643	1.6550	1.4849
1042	1.6428	1.4799

In PLM, light emerging from the lamp passes through the first filter (polarizer) before following the usual path to the sample. This light is plane polarized as shown in Figure 7.1. After passing through the sample, the emergent light passes through the second (analyzer) filter, which is oriented perpendicular to the polarizer. This is referred to as the **crossed polar** orientation. If there is no sample on the stage, the field of view will appear completely black because all light is blocked. The same applies if an anisotropic material is placed in the light path.

An image will be visible only if a sample interacts with polarized light in such a way as to prevent the analyzer from filtering all the emergent light out before it reaches the eyepiece. This is what occurs with birefringent materials, as shown in Figure 7.8, where polarized light enters the sample and splits, and one vibrational direction travels faster than the other. This creates the slow (retarded) and fast components. As shown in the figure, the split beam continues to propagate in the same direction, with one component vibrating perpendicularly to the other. The emergent light is recombined at the analyzer. If no retardation occurs, the recombined light will not interfere and all light will be blocked by the analyzers. However, if retardation occurs, the recombined beams will interfere and some wavelengths of light will be able to pass through the analyzer because their orientation is now different from what it would have been if no sample was present or an anisotropic sample was in the beam. Relative retardation can be described mathematically as

$$\Gamma = \left(n_e - n_o \right) t \qquad (7\text{-}3)$$

where t is sample thickness. The slow axis is the one in which the orientation of atoms and molecules is such that the refractive index is the higher of the two. The fast axis is the direction in which the refractive index is the lower. In the O and E parlance, samples with positive birefringence, $n_e > n_o$, the extraordinary E ray

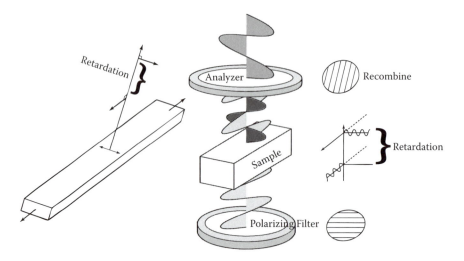

FIGURE 7.8   Light path in polarizing light microscope.

is the slower. The opposite is true for negatively birefringent materials. When the fast and slow rays recombine at the analyzer, interference results and interference colors are visible. The magnitude of the difference in propagation speed of the fast ray relative to the slow ray, and the thickness of the sample, determine the degree of retardation and thus the interference colors produced.

At this point, we should expand our terminology to better describe the waveform that emerges from a birefringent material. Figure 7.8 is a simplification of what actually occurs. In the case of calcite, the optic axis was parallel to the incident beam and the split beams traveled in different directions and emerged in physically separated places on the crystal. In the case of a sample placed on a microscope slide, the optic axis is usually perpendicular to the incident beam as shown in frame (c) of Figure 7.7. In this case, the split beams follow the same trajectory but vibrate perpendicularly to each other, one in the fast direction, and one in the slow. The slow component is retarded relative to the fast component, and the thicker the sample, the greater the retardation. These waves are superimposed on each other at the analyzer to create interference colors. However, the colors are not created by simple constructive or destructive interference since the two rays are not vibrating in the same plane.

One way to understand how the fast and slow rays interact is to combine them using vector addition as discussed previously (Chapter 1, Section 1.2) and depicted in Figure 7.9. At any point along the trajectory, the locations of the O and E rays can be described by x, y, and z coordinates, as shown in the waveform to the left. We defined the plane of the paper as the north–south axis (N-S) and the plane perpendicular to the paper as the E-W axis, a notation convention frequently used with PLM. For clarity in Figure 7.9, *east* is spelled out to differentiate it from the E (extraordinary) component.

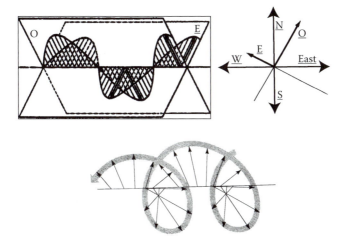

FIGURE 7.9 How light recombines at the analyzer. The top frame shows how vector notation is used to combine the parallel and perpendicular components. The lower frame shows the path traced out in space. The resulting cross-section is elliptical.

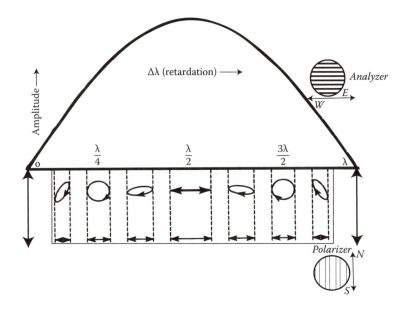

FIGURE 7.10 Elliptical polarization as a function of retardation. The amplitude of the reconstucted wave is maximized when the retardation is equal to half a wavelength.

At any point along the axis of propagation, the split beam can be described in terms of vector notation and vector addition, which is essentially what happens when the beams recombine. If you were to stand facing the oncoming split beam, you would see the components of the vector as shown in Figure 7.9 (top right). These two vectors can be combined by simple vector addition. If this is done for every point along the trajectory of travel, the additive vectors will sweep out a path in space that is defined by the degree of retardation of the slow beam relative to the fast beam as shown in the bottom frame of Figure 7.9. At certain points, the cross-section of the corkscrew pattern is elliptical, some points circular, and some points linear.

The degree of retardation of the O and E (fast and slow) components dictates what the cross-section looks like and thus what type of interference occurs when the beams recombine (Figure 7.10). If no retardation (propagation in the fast direction = propagation in the slow) occurs, the added vector traces a line. If the offset is equal to $\lambda/4$, the path traced is a circle. Most samples result in elliptically polarized combined rays as shown in Figure 7.10 (bottom).

Based on this depiction, we can predict what will occur in a polarizing microscope under the crossed polar condition. Plane polarized light exits the polarizer and interacts with the sample. If the sample is birefringent, the beams split into parallel and perpendicular components and retardation occurs as described by Equation 7-3. After light emerges from the sample (Figure 7.8), it travels to the analyzer where the components recombine. The intensity of the light passed through the analyzer depends on the retardation, with maximum intensity occurring when the retardation between the fast and slow ray is equal to half a wavelength. The reason for this can be understood by the vector addition shown in Figure 7.9. At half a wavelength difference, the crest of the wave in the xy plane is at maximum while the wave in the yz plane is at minimum (zero), leaving only the xy component, the one parallel to the analyzers.

7.5   INTERFERENCE COLORS AND EXTINCTION POINTS

Interference colors arise because the light emitted from a lamp is polychromatic and each wavelength experiences a different degree of retardation based on the wavelength-dependent nature of refractive indices. This provides the opportunity for constructive and destructive interference to occur at the analyzer. Some wavelengths are removed entirely while others experience constructive interference and are reinforced. A highly colored, high contrast image can be created from this combination of factors: intrinsic sample birefringence, thickness of sample, and orientation of the polarizer and analyzer. Interpretation of the colors can tell a skilled microscopist a great deal about a sample.

Figure 7.11 illustrates how multiple wavelengths of light combine to form new colors. Six representative wavelengths are shown separately with the combined pattern shown at the top. At any point on the top curve, the color results from a linear combination of the individual wavelengths. The white light at the apex is seen when roughly equal contributions are seen from all colors. Smaller and nearly equal contributions result in a gray color. Following the thick gray line with the circles, you can see that the violet and blue waves are just past their apex, the red, orange, and yellow waves are just short of their apex, and green is at the apex. Combining these nearly maximized waves results in observation of a white color. This process is a reasonable approximation of what occurs at the analyzer of a polarizing microscope and explains the interference colors observed. Other combinations lead to perceptions of different colors. Note the three shaded boxes in the top frame. These correspond to regions of first, second, and third order interference colors. The first order colors are always the most intense.

The most common tool for interpreting interference colors and birefringence is the **Michel-Levy chart** (Figure 7.12). Although the figure is not in color, color versions are available from microscope vendors and on the Internet.* Using the chart is simple. By using the observed color and one variable (thickness or birefringence), the other variable can be determined by following the diagonal line. Typically, thickness is known (as discussed in Chapter 5) and the birefringence is unknown. When looking at the colored version, you will note that most of the space to the left side is gray or white, corresponding to the least retardation and the left-most portion of Figure 7.11. The first order colors are most intense and

---

* Example websites current as of July 2009:
  http://www.olympusmicro.com/primer/java/polarizedlight/michellevy/index.html
  http://www.microscopyu.com/articles/polarized/michel-levy.html

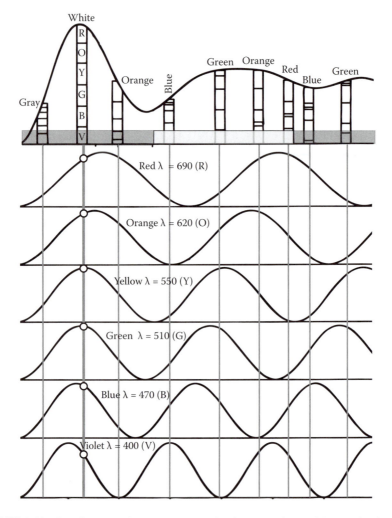

FIGURE 7.11 Interference colors seen as a result of constructive and destructive interference. The light boxes in the top frame correspond to first, second, and third order interference colors (left to right).

fade noticeably across the chart until most of the colors in the fifth and sixth order regions are washed-out yellows and pinks.

After you obtain a birefringence value using a table, a mineral or fiber may be tentatively identified. With experience, you will be able to visualize cross-sections of samples such as fibers using the patterns of interference colors as shown in Figure 7.13. Notice the relatively smooth gradient from light at the edges to dark in the center of both fibers; this corresponds to a roughly circular cross-section. Note also the cross-shape where the fibers intersect. The fibers are perpendicular to each other and at maximum brightness, light from the polarizer is passing through both

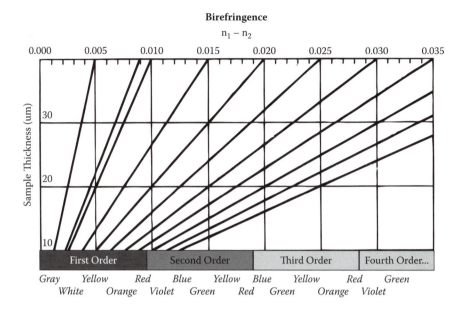

Birefringence

$n_1 - n_2$

FIGURE 7.12   Portion of Michel-Levy chart shown in black and white. Retardation is approximately 100 to 500 nm for the first order region, 560 to 1150 nm for the second order, 1160 to 1700 nm for the third order, and 1700 to 2200 nm for the fourth.

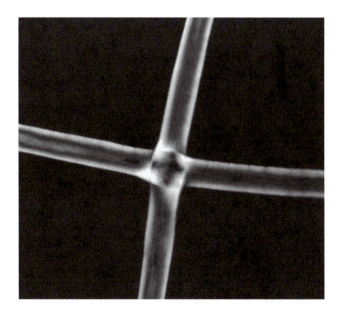

FIGURE 7.13   Two nylon fibers under crossed polars.

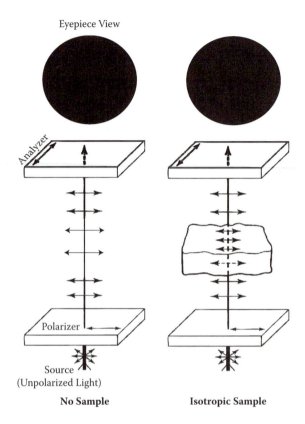

Eyepiece View

No Sample   **Isotropic Sample**

FIGURE 7.14   Selected eyepiece views from polarizing light microscopy.

fibers. The subjective nature of judging color should be kept in mind when you use numerical values for birefringence (or thickness, if this is the variable of interest).

Finally, when we consider interference colors and microscopic samples in PLM, one other optical characteristic is of interest. Figures such as 7.8 show a sample in the light path with the optic axis in a fixed position. However, stages and samples rotate, and on any slide may reveal many crystals at different orientations. Thus, it is important to rotate the slide and observe when interference colors are at the maximum and minimum (extinguished completely). First, consider the situation depicted in Figure 7.14. With no sample in the path and the polarizer and analyzer crossed, no light reaches the eyepiece because all light is blocked. Similarly, when an isotropic sample such as table salt is placed in the light path, all the light is blocked because there is no birefringence, beam splitting, retardation, or interference.

In Figure 7.15, a birefringent sample is now in the light path. As the sample is rotated on the stage, there will be four points of **extinction** in which all light is blocked. To understand why, return to Figure 7.9 and the vector notation. As the sample is rotated, the orientation of the O and E rays to the right of the figure changes while the north–south and east–west orientations of the polarizer and

Anisotropic Crystal

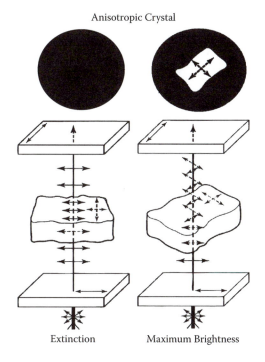

Extinction   Maximum Brightness

FIGURE 7.15 Eyepiece views of a birefringent crystal as a function of sample orientation.

analyzer do not. At four points of the rotation, the vibrational direction in the crystal will be perpendicular to the analyzer and as a result, no light will be transmitted. At points between, the intensity (but not the color) varies until a maximum intensity is reached as shown in Figure 7.15. At that rotation and angle, the two vibrational directions in the crystal are at 45° to the analyzer and when added, the components are parallel to the analyzer and all the light can pass to the eyepiece. **Extinction angles** are used to identify different types of crystals; a detailed discussion of this is beyond the scope of this text. Consult the references at the end of the chapter for more information on this topic.

7.6 RETARDATION PLATES AND COMPENSATORS

A material such as cellophane tape (thin and pseudocrystalline) is invisible when placed parallel to the analyzer. When a sample is rotated to an angle of 45° relative to the analyzer (also 45° off the optic axis), the tape appears as a bright but essentially gray image. Why is it not highly colored? Recall that retardation is a function of sample thickness and birefringence (Equation 7-3). In the case of the tape, the magnitude of the retardation is small and the sample is very thin. As a result, no wavelengths in the visible range are completely blocked. Several factors contribute to the type of interference colors seen. In some cases, the lack of characteristic colors limits the analytical capability of PLM.

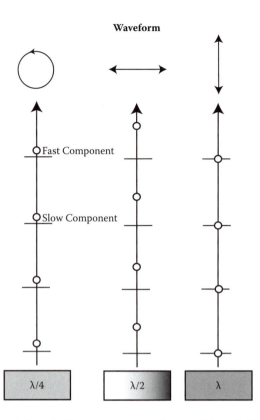

FIGURE 7.16   Retardation of mid-range green wavelength of light. Refer to Figure 7.10 for information regarding waveform.

One approach to enhance color and contrast is to use thicker samples, but in microscopic work, this variable is usually beyond our control. The relative retardation of the fast and slow rays is a function of the sample, so the only other approach to enhancing color and contrast is to somehow enhance or magnify the retardation. This can be accomplished using retardation filters inserted into the light path after the sample and before the analyzer. The function of a retarder is to add a constant and known amount of retardation to that generated in the sample. Quarter waveplates introduce a constant retardation of $\lambda/4$; a half-wave plate, $\lambda/2$; and a full waveplate creates one wavelength of additional retardation (Figure 7.16). The full waveplate is sometimes called a first order red plate because the field of view under crossed polars, with a compensator in place, is magenta rather than almost black because of retardation, compensation, and interference.

The midpoint of the white light spectrum is about 550 nm and a full waveplate generates single wavelength retardation at this midpoint (Figure 7.16, far right). Light in the midrange will emerge from the compensator more or less linearly polarized and will be blocked by the analyzer. Other wavelengths exhibit different retardations and emerge from the compensator elliptically polarized as shown in

Figure 7.10, with different intensities reaching the analyzer. In effect, the light that reaches the eyepiece is white light minus the midrange green contribution, leaving the characteristic magenta seen as background when a full waveplate is used. When a birefringent sample is placed in the light path, the retardation effects are amplified by the waveplate.

7.7   EXAMPLE APPLICATIONS

While our discussion has focused on minerals and crystalline solids, PLM is suitable for many applications beyond mineralogy and chemical microscopy. These include applications in materials science (polymers) and medicine. Also, PLM can be performed in reflecting mode as well as in transmittance, opening even more applications for the technique. We will look at a few examples.

Cellophane tape (cited in an earlier example) illustrates how PLM can be applied to materials that lack the typical rigid chemical crystal structure seen in calcite. It is also used in one of the exercises at the end of this chapter. It is a polymer derived from cellulose and was one of the first commercial polymers. Cellophane and many other polymers are considered pseudocrystalline materials—they exhibit regularities in structure and differences in refractive indices that are functions of direction of propagation. The directionality and fast and slow orientations are obvious; the slow axis is along the length of the tape and the fast axis is across the width. Many synthetic fibers such as nylon follow this general pattern. Cotton fibers under PLM conditions show patches of interference colors but no consistent pattern, corresponding to partially ordered regions in the cellulosic matrix. Rayon, an early semi-synthetic fiber, consists of reconstituted cellulose. The process of drawing the fiber contributes to the order and pseudocrystalline nature of rayon, resulting in birefringence and interference colors.

The fibers shown in Figure 7.17 illustrate some of the powerful characterization capabilities inherent in PLM. Although the image is shown in grayscale, you can still use the light and dark band patterns to visualize the cross-sectional characteristics. The twist in the fiber in the upper right corner further helps visualize the cross-section. Using a Michel-Levy chart and the interference colors, you could easily estimate the thickness of the fiber in different zones (edge to middle). Finally, in the fiber to the left, you can barely distinguish dark specks on the surface; these are likely delustering particles used to induce the light impinging on the fiber to scatter, dulling the finish.

Starch grains are somewhat nondescript when viewed under typical light microscopic illumination but illustrate another feature of polarized light and interference patterns. Under crossed polars, black crosses appear in the middle of starch grains (Figure 7.18). No matter how the grain is rotated, the dark pattern remains at the same orientation because starch is composed of an orderly biopolymeric form of glucose. A similar pattern was seen in Figure 7.13 where the two fibers crossed. Such interference patterns can be used to obtain much more information about crystal structures and habits, discussed in the next chapter.

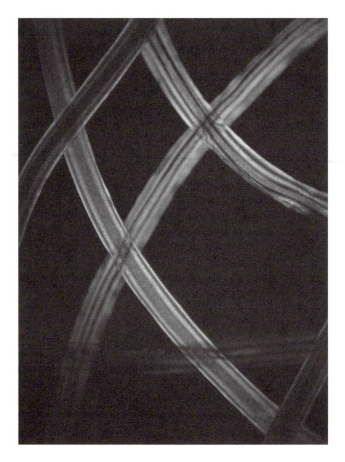

FIGURE 7.17   Fibers under crossed polars.

FIGURE 7.18 Starch grains seen under crossed polars.

REFERENCES AND FURTHER READING

F.M. Ernsberger, *Polarized light in glass research*. PPG Industries, 1970.

N.H. Hartshorne and A. Stuart, *Crystals and the polarizing microscope*, 4th ed., American Elsevier, 1970.

D.S. Kliger, J.W. Lewis, and C.E. Randall, *Polarized light in optics and spectroscopy*, Academic Press, 1990.

D.B. Murphy, Polarization microscopy, in *Fundamentals of light microscopy and electronic imaging*, John Wiley & Sons, 2001.

D.B. Murphy, Properties of polarized light, in *Fundamentals of light microscopy and electronic imaging*, John Wiley & Sons, 2001.

Nikon Corporation, *Michel-Levy interference chart* [cited November 23, 2008]. Available from http://www.microscopyu.com/articles/polarized/michel-levy.html.

E.M. Slayter and H.S. Slayter, Polarized light, in *Light and electron microscopy*, Cambridge University Press, 1992.

E.M. Slayter and H.S. Slayter, Wave interactions, in *Light and electron microscopy*, Cambridge University Press, 1992.

R.E. Stoiber and S.A. Morse, *Microscopic identification of crystals*, Robert E. Krieger, 1981.

U.S. Naval Research Laboratory, *Crystal lattice structures* [cited November 23, 2008]. Available from http://cst-www.nrl.navy.mil/lattice/.

QUESTIONS

1. Explain the trend seen in Table 7.1. Why does refractive index change as a function of wavelength?
2. Why is birefringence a wavelength-dependent phenomenon?
3. On a molecular level, explain why a beam of polarized light propagating down the optical axis of a calcite crystal is **not** split.
4. What would the equivalent image for Figure 7.5 look like for an isotropic crystal such as NaCl?
5. When cotton fibers are viewed under crossed polars, some regions show interference colors and some do not. Explain on a molecular level why this is so.
6. Some commercial fibers have cross-sections that look like dumb bells. How would they appear under crossed polars, assuming the fibers are birefringent?

EXERCISES

Exercise 7.1: Polarizing films

Materials needed:*

　　Strong overhead lighting
　　Light table or other intense light source that illuminates from below (large flash-
　　　　light with lens approximately the size of the polarizing filters works well)

---

* These supplies are available from many educational warehouses.

Linear polarizing filters (available from Edmunds Scientific, http://scientific-sonline.com/, Wardsci.co, or Teachersource.com)
Half and quarter wave retardation filters (same sources)
Calcite crystal (same sources and others)
Computer monitor with white background

This series of exercises requires reasonably clear calcite and polarizing filters. Good lighting from above and below is also needed for best results. The exercise represent a macro version of what occurs in a PLM. Note that some commercial calcite blocks are too thick and cloudy to see through; if this is the case, you can use a hammer and chisel to flake off thinner clearer sections.

For the first experiment, you can use a computer monitor as a light source. Hold a polarizing filter against a white background (such as a blank word processing document) with the screen intensity at full brightness. Rotate the filter and note observations. Is the light coming from the monitor polarized or not? Does this make sense based on how a monitor works? HINT: Do a little research on your monitor type and find out how the light is generated.

Next, take the calcite crystal or section of it and place it over a typed document. Provide sufficient lighting from above and make sure the crystal is transparent enough to see through. You should see double images of each letter, one produced by illumination from the E ray and one produced by the O ray. Rotate the crystal and note your observations. If you have different thicknesses of calcite, try viewing them and note what you observe. Can you identify the direction of the optic axis? (Refer to Figure 7.6.)

Tape one of the polarizing filters over the front of an intense flashlight so that the filter stays in place. Allow some of the tape to project into the lighted area. Place the second polarizing filter on top of it and rotate it. Note where total extinction occurs and the color of the background. What about the cellophane tape? Does it ever disappear completely? Next, place a piece of calcite on the lower filter (simulating a polarizer in a microscope) and cover with the second filter (simulating an analyzer). Rotate the top filter without disturbing the calcite. When the background is completely dark, what does the calcite look like?

If you can obtain a retardation film, place it directly over the polarizer and then place the analyzer film atop it. Rotate the analyzer and note what you see. How has the background color changed at crossed polar conditions? Explain, based on the discussion in this chapter. Finally, place two strips of cellophane tape on the retardation filter, one crossing the other at a right angle. Cover with the analyzer film. Rotate the analyzer and note your observations, then rotate the film with the tape and note your observations. What do you see where the two tape samples cross and why?

Exercise 7.2: Polarizing light microscopy and use of Michel-Levy chart

Materials needed (same as for Exercise 5.1, Micrometry):

Polarizing light microscope and slides
Michel-Levy chart (available from vendors or educational supply stores)
Calcite flakes
Corn starch

Nylon fibers
Rayon fibers
Cotton fibers
Cellophane tape on a slide with torn edges visible

Make sure your microscope is clean and aligned for *Kohler Illumination* (Chapter 4). Optimal PLM results and photography require optimal illumination. If you completed Exercise 5.1, you will already have the prepared slides and data needed for the Michel-Levy chart part of this exercise. If not, perform Exercise 5.1 now. After you obtain the data, you can begin. Observe each sample under polarized conditions and then under crossed polars. Describe in detail what you observe as you rotate the samples on the stage. Make sure you can explain your observations based on material discussed in this chapter. If you have access to waveplates, insert them and repeat the procedure. Note the different colors of the background at crossed polars.

Some of these samples (e.g., starch grains and cotton) are not appropriate for Michel-Levy applications. Why? Can you explain the appearance of the cotton based on what cotton is and how cotton fibers are produced? What explains the appearance of the tape? Can you identify the fast and slow axes? What causes the pattern you see in the starch grain? Compare this to Figure 7.13 and comment on the similarities and differences.

Use the Michel-Levy chart, the sample thickness, and the interference colors observed to measure the birefringence values of the remaining samples. Can you tell what the cross-section of the fibers is? As with the micrometry exercises, it is important to analyze several samples with different orientations on the slide to obtain a reasonable estimate of birefringence.

8 Basic Crystallography

The imaging of particles requires an understanding of the forms that substances may take. These forms fall within a number of defined classes for crystalline substances. As noted in Chapter 7, crystal order and structure dramatically affect the way in which polarized light interacts with a substance. We touched upon some basics in that chapter and will now discuss crystal structure in more detail and explain how microscopy (PLM in particular) can be utilized to probe crystal morphology. Apart from glass, which is an amorphous substance (lacking a finite crystal structure), a microscopist will encounter many crystalline forms. These morphological properties provide valuable tools for identifying samples.

8.1 CRYSTAL SYSTEMS

A crystal can be envisioned as a combination of atoms, ions, or molecules. Since a crystal has a regular structure, a basic unit known as the **unit cell** of the crystal lattice provides symmetry (Figure 8.1). A crystal is formed by combining a number of unit cells. Studies of crystal forms revealed 7 basic systems divided into 32 classes. Table 8.1 lists the crystal systems. A crystal may grow at different rates at each of its faces and the resultant crystal may not display the expected symmetry. The crystal will, however, maintain the overall symmetry dictated by its internal organization. Figures 8.2 through 8.6 depict hexagonal, orthorhombic, triclinic, and monoclinic crystals.

8.2 ISOTROPIC AND ANISOTROPIC CRYSTALS

Crystals can be divided into two main classes: those with a single RI (**isotropic**) and those with more than one RI (**anisotropic**) . This was discussed in the previous chapter and now we can extend that knowledge. The anisotropic crystal can be further subdivided; crystals with two refractive indices are called **uniaxial**; those with three are known as **biaxial crystals**.

8.2.1 ISOTROPIC CRYSTALS

An **indicatrix** is a diagram that can help us understand the combination of refractive indices in a crystal. Imagine a solid body on a system of three axes. Each axis represents the RI of the crystal in that direction. An isotropic substance, for example, has one refractive index regardless of orientation. Thus, as a solid body, this can be represented on our axes as a sphere around the origin. Why such a shape? All light passing through the crystal is retarded to the same extent according to **Snell's law** (Eq. 2.1) (the local environment is the same; see the description of the unit cell). The indicatrix can be constructed as a two-dimensional representation. This is simple for

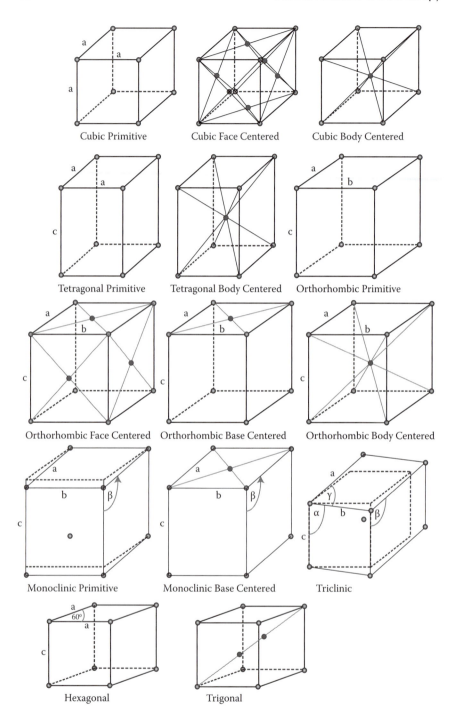

FIGURE 8.1 Bravais system of crystal lattices. The side lengths are indicated by *a*, *b*, and *c*. All angles are 90° unless they are marked α, β, or γ.

TABLE 8.1

Crystal Systems and Bravais Lattices Indicating Unit Cell Descriptions and Refractive Indices

Crystal System	Unit Cell Description	Bravais Lattice	Refractive Indices
Triclinic	All three axes of unequal length, none perpendicular	Primitive	Three; two optic axes unrelated to crystallographic axes and rarely perpendicular to crystal face
Monoclinic	One axis perpendicular to two at obtuse angle	Primitive, base centered	Three; two RIs mutually perpendicular in plane of a and c; third parallel to b
Orthorhombic	All three axes of unequal length, all perpendicular	Primitive, base centered, body centered, face centered	Three; two optic axes, not parallel to crystallographic axes, lying in plane of two of them; only two of three RI exhibited in any principal view; only intermediate values in other views; β shown only if light travels parallel to optic axis
Rhombohedral (or Trigonal)	All three axes of equal length, none intersecting at equal angles, may be considered part of hexagonal system	Rhombohedral (two centered elements)	Two; see hexagonal system
Tetragonal	Two axes of equal length, all perpendicular	Primitive, body centered	Two; optic axis parallel to c axis; crystals appear isotropic if light travels in this direction; two RIs when light travels perpendicular to c axis; ε parallel to c and ω perpendicular to c; isotropic views give values of ω only
Hexagonal	Three planar axes of equal length intersecting at equal angles, an axis perpendicular to this plane	Primitive	Two; optic axis parallel to c axis; crystals appear isotropic if light travels in this direction; two RIs when light travels perpendicular to c axis; ε parallel to c and ω perpendicular to c; isotropic views give values of ω only
Cubic (or Regular)	All three axes are of equal length, all of the axes are perpendicular	Primitive, body centered, face centered	One; optically isotropic; all orientations alike; one RI that is independent of direction of light

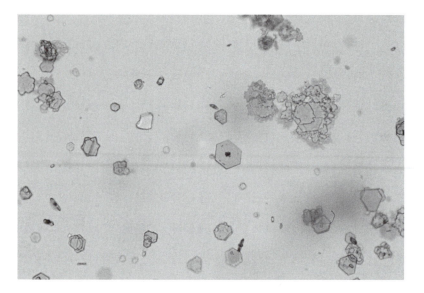

FIGURE 8.2 Reaction of lead nitrate [Pb(NO$_3$)$_2$] and potassium iodide [KI] to form lead iodide [PbI$_2$] (hexagonal), 40× objective.

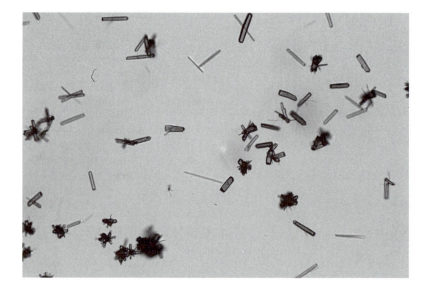

FIGURE 8.3 Reaction of cobalt acetate [Co(CH$_3$CO$_2$)$_2$] and potassium mercuric thiocyanate [K$_2$Hg(SCN)$_4$] to form [Co(SCN)$_2$Hg(SCN)$_2$] (orthorhombic). This product is very soluble and only forms after evaporation of some of the water. The intense blue color of the product indicates that the Co^{2+} is in a tetrahedral environment.

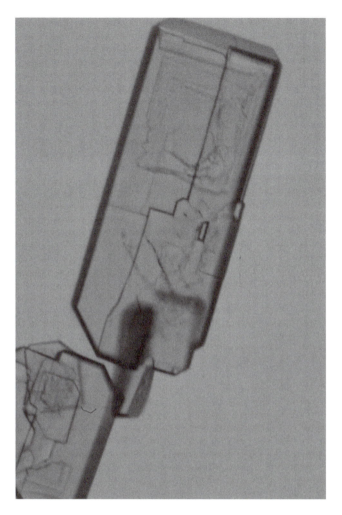

FIGURE 8.4 Recrystallized potassium dichromate [$K_2Cr_2O_7$] (triclinic).

an isotropic substance because a sphere will reduce to a circle in two dimensions; regardless of the direction, the RI remains constant.

8.2.2 UNIAXIAL CRYSTALS

In a tetragonal crystal, two of the three refractive indices are equal (Figure 8.7). The main axis is called the unique direction since it is the only axis of its size. Perpendicular to the unique axis are two equal axes. The orientation of the atoms in the plane of these two axes is the same. Thus an isotropic environment is encountered by light traveling down the main axis. This ray is called the **ordinary ray** or omega (ω). The other ray (two equivalent rays in a uniaxial crystal) is called the **extraordinary ray** or epsilon (ε). See Figure 8.8. We saw this notation in Chapter 7 in which O and E served as the representative symbols.

FIGURE 8.5   Recrystallized copper acetate [Cu(CH$_3$CO$_2$)$_2$] (monoclinic).

The indicatrix for a uniaxial crystal can be described using an ellipse. The orientation of a crystal on the stage of a microscope (where the light is plane polarized) is very important for evaluating RIs. Consider a tetragonal crystal positioned on a stage where the main axis is parallel to the optical path. Light moving up through the crystal will encounter an isotropic environment. If the crystal is orientated so that one of the equivalent axes is parallel to the optical path, the anisotropic nature of the crystal can be evaluated.

When a crystal is mounted in a medium whose RI is unequal to either of its own RIs (as per Chapter 5), the following situation is encountered. If, for example, the ordinary (ω) ray is parallel to the plane of polarization and the extraordinary (ε) rays are perpendicular to this plane, the crystal will be visible at all rotations of the stage.

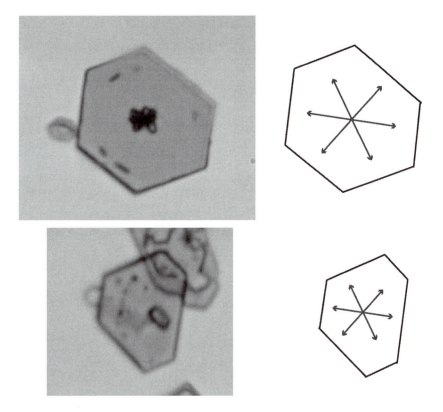

FIGURE 8.6 Hexagonal crystals of lead iodide [PbI$_2$]. Notice that they are not regular hexagons, but differential growth occurs at each crystal face.

When the extraordinary axis is parallel to the plane of polarization, ε is observed; at a 90° rotation, ω is observed; between these two points, ε' is observed (ε' lies between ω and ε).

ε' may be observed in other situations. If the extraordinary ray is in the plane of polarization and the plane of the ordinary rays is not parallel to the microscope axis (offset at some angle θ), the extraordinary ray "projected" onto the stage is ε' ($\varepsilon' = \varepsilon \cdot cos\theta$). Thus, the only time when ε may be observed is when the optical axis of the uniaxial crystal is perpendicular to the microscope axis.

If the RI of the mounting medium is changed to or near the RI of the extraordinary ray, rotation of the stage will result in disappearance of the crystal at two orientations 180° apart. When the stage is rotated 90° to one of these extinction points and the refractive index of the mounting medium is changed so that the crystal disappears, the RI of the ordinary ray equals the RI of the mounting medium.

For an **optically positive crystal** (+), $\varepsilon > \omega$, and for an **optically negative crystal** (–), $\varepsilon < \omega$.

The indicatrix is a three-dimensional representation of the interrelationship of a crystal's RIs. The indicatrix of a uniaxial crystal is an **ellipsoid of revolution** where

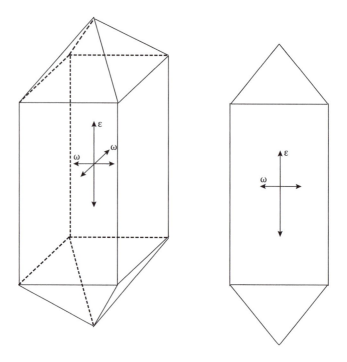

FIGURE 8.7   Refractive indices of a tetragonal crystal (uniaxial).

the extraordinary ray is represented by the **major axis** and the ordinary ray by the **minor axis** for an optically positive crystal. If you look down the major axis, the projected area of the indicatrix is a circle that represents the local isotropic nature of the crystal down this axis. When you look down the minor axis, the projected length of the major axis equals ε. If the indicatrix is tilted at any other angle, the projected length of the ordinary axis always represents ω and the major axis will always represent a value of ε' (always less than ε). A similar situation applies to an optically negative indicatrix. Remember that the indicatrix for an optically negative uniaxial crystal is not simply the rotation of that of an optically positive crystal. This information applies to hexagonal crystals except that they exhibit three ordinary rays that are equal.

8.2.3   BIAXIAL CRYSTALS

Crystals from the orthorhombic, monoclinic, and triclinic systems are biaxial. They have three distinct RIs because the local environments of all three axes of the unit cell of the crystal differ. These three refractive indices are represented by α, β, and γ (where α < β < γ).

Since each of the refractive indices of biaxial crystals is different, the indicatrix of a biaxial crystal is a **triaxial ellipsoid**, rather than the ellipsoid of revolution of a uniaxial crystal (Figure 8.9). In the ellipsoid of revolution the ω plane was circular. The "equivalent" plane for a biaxial crystal is the α – β plane (which in itself is

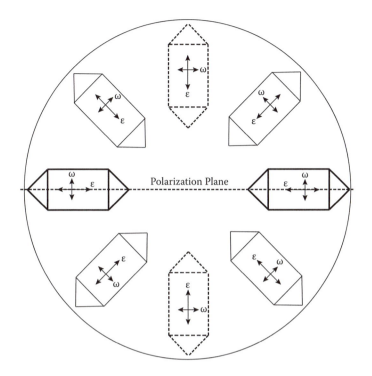

FIGURE 8.8 Example of tetragonal crystal (uniaxial) mounted in a medium with a refractive index close to that of the ordinary ray (ω). The ordinary ray runs down the length of the crystal. The crystal is oriented such that the third refractive index (ε) is parallel to the optical axis of the microscope (you are looking down this axis). When the extraordinary ray is parallel to the plane of polarization, its refractive index (ε) can be determined. Upon rotation of 90°, the ordinary ray is parallel to the plane of polarization, but because the mounting medium is the same (or similar) the crystal will disappear. By changing the mounting medium, the ordinary ray can be determined.

ellipsoidal). A projected area that rotates along the β axis such that α tends to γ will result in a value of α' = β. This area is circular in shape and a line perpendicular to this circle (though the center of the indicatrix) represents one of the optical axes of the crystal (remember that the extraordinary ray in the uniaxial indicatrix was also perpendicular to this plane). If the tilting were to take place in the opposite direction, an equivalent circle would be found.

In a similar manner, the second optical axis can be determined. These two optical axes form an acute angle called the **optic axis angle** (2V). Since the rotation to find the optical axes is around the β axis, both of the optical axes, and thus the optic axis angle, are found in the α – γ plane. The optic axis angle is specific to particular substances and can aid in their identification.

Remember from the discussion of the uniaxial indicatrix that the rays traveling down the optic axis (ε) experience an isotropic neighborhood. The same is true for biaxial crystals (light traveling down the optic axes experiences isotropic

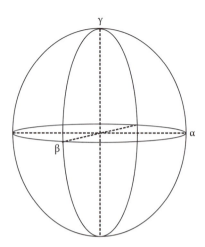

FIGURE 8.9   Refractive indices of a biaxial crystal can be illustrated by a triaxial ellipsoid.

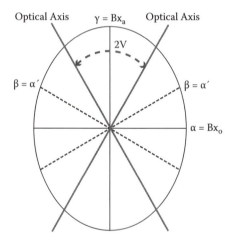

FIGURE 8.10   Indicatrix of optically positive biaxial crystal.

neighborhoods). For a biaxial crystal, if β is closer to α than it is to γ, the crystal is optically positive (+). See Figure 8.10.

From these principles it is easy to construct a biaxial indicatrix. Start with an indicatrix for an optically positive crystal. We know that $\alpha < \beta < \gamma$, so it is possible to indicate the value of β on the $\alpha - \gamma$ plane. In an optically positive crystal, β is closer to α than it is to γ and β is thus represented by α'. The optical axes are perpendicular to β. Note for the positive figure that 2V is bisected by γ, since β is closer to α. Do the same for the negative rotation. Remember for a negative indicatrix, β is closer to γ and thus 2V will be bisected by α.

It is important to note that when the β axis is not perfectly aligned with the plane of polarization, the value of α' and γ' (where $\alpha' < \alpha$ and $\gamma' < \gamma$) will be obtained. Thus the orientation of the crystals on the stage of the microscope plays a big role. If we assume that the orientations are random, rotation of the stage will cause changes in the images at different angles. For a biaxial crystal, you can determine the values of α and γ by varying the RI of the mounting medium.

For an unknown crystal, the general approach is to evaluate the crystals for their refractive indices by variation of the mounting media. If for a particular mounting medium all the crystals disappear, they must be isotropic. If all of the crystals disappear at two positions during rotation of the stage (180° apart), the crystal must be uniaxial. If only a few of the crystals reach extinction during rotation of the stage, the crystal is biaxial. Also remember that we are dealing with a single mineral. A mixture of isotropic, uniaxial, and biaxial crystals is far more complex and beyond the scope of this introductory text. However, you now have the basic knowledge needed to explore these advanced topics. Consult the references in Appendix 4 as a starting point.

REFERENCES AND FURTHER READING

V. Agrawal, G. Chadha, and G. Trigunayat, *Acta Crystallographica*, A26: 140–144, 1970.

P. Atkins, *Physical Chemistry*, Oxford University Press, 1983.

B. Bleaney and K. Bowers, *Proceedings of the Royal Society of London. Series A,* 214: 451–465, 1952.

E. Chamot and C. Mason, *Handbook of chemical microscopy,* Vol. II, John Wiley & Sons, 1948.

E. Chamot and C. Mason, *Handbook of chemical microscopy,* Vol. I, John Wiley & Sons, 1948.

E. Dana, *A textbook of mineralogy*, John Wiley & Sons, 1932.

J. Delly, *Microscope* 37: 139–166, 1989.

J. Jeffery and K. Rose, *Acta Crystallographica*, B24: 653–662, 1968.

A. Lide, Ed., *CRC Handbook of chemistry and physics*, CRC Press, 1983.

QUESTIONS

1. Describe the refractive indices in isotropic and anisotropic substances.
2. Describe how you would determine the refractive index of ω for an optically negative uniaxial crystal.
3. Describe what you would see if the crystal described in question 2 is mounted in a medium that has an RI between ε and ω and the stage is rotated.
4. If only a few crystals reach extinction during rotation of the stage, an unknown crystal must be biaxial. Why is this so?
5. A triclinic crystal is rotated on a stage with the plane of polarization in the west–east direction. When will the crystal reach extinction?

EXERCISES

Exercise 8.1: Indicatrix for negative biaxial crystal

Materials needed:

Pencil
Paper

Construct an indicatrix for an optically negative biaxial crystal via the process described in the chapter. Remember that for a biaxial crystal, if β is closer to γ than it is to α, the crystal is optically negative (–).

Exercise 8.2: Optical signs of biaxial substances

Materials needed:

Pencil
Paper

Designate the refractive indices (RIs) as α, β, and γ for each of the compounds listed in the table below, then calculate the values of β – α and γ – β to determine the optical signs of the compounds.

Compound	Crystal System	RI 1	RI 2	RI 3
Ammonium cobalt sulfate	Monoclinic	1.490	1.492	1.503
Manganese(II) *meta*silicate	Triclinic	1.733	1.740	1.744
Potassium nitrate	Orthorhombic	1.335	1.5056	1.5064
Calcium *meta*silicate	Monoclinic	1.616	1.629	1.631
Potassium sodium tartrate	Orthorhombic	1.492	1.493	1.493

9 Chemical Microscopy

Chemical microscopy utilizes the techniques of qualitative analysis (organic and inorganic) to study samples. When these techniques were first used, many were considered definitive based on nothing more than observation of crystal habit, optical properties, and appearance under a microscope. Although this is no longer the case, microchemical analysis can be a powerful first step for many chemical determinations. The single greatest advantage of microchemical techniques over their macro counterparts is the tiny amount of sample consumed.

Additionally, when solids are formed, their morphology and behavior under polarized light conditions add significantly to the information about a sample. For example, you could apply the knowledge gathered from the crystallography techniques in Chapter 8 to help characterize a sample. The challenge of microchemistry is mastering the methods and techniques required to perform accurate, precise, and reliable testing. One advantage of microscopy is the relative ease of combining physical and chemical testing and the ability to work with molecular and ionic compounds.

9.1 PHYSICAL AND CHEMICAL PROPERTIES

In any characterization process, we can measure physical and chemical properties. Most of what has been covered so far relates to physical properties such as morphology and color. A physical property is one that can be measured or observed without causing a chemical change. For example, the melting point of a solid is a physical property and measuring it does not change the substance. Ice, water, and steam consist chemically of H_2O; melting ice changes only the physical state, not the H_2O composition. Other physical properties include density, color, refractive index (RI), and optical properties discussed in previous chapters.

Chemical properties can be measured only by invoking a change in the substance of interest. The flammability of octane (C_8H_{18}) is measured by combusting octane—a decomposition that results in the creation of carbon dioxide and water (ideally). Other chemical properties include corrosivity and reactivity. Using microchemical methods, we can measure both physical and chemical properties of substances and products of chemical reactions performed in the course of microchemical testing.

Some bench-scale methods such as solubility testing, melting point determination, and vaporization and evaporation studies are easily reducible to the microscopic scale. With practice, you may find microscopic studies easier than the larger scale equivalents because of the ease of observation of the results under a microscope. Other methods are more challenging because of the micromanipulation needed and the tendency to use too much reagent or sample. Microchemical skills come with practice and once mastered are generalizable to a variety of testing applications.

9.2   PRECIPITATES, SOLIDS, AND SOLUBILITY

Solids and their formation on a microscope slide involve the same chemistry as form-ing a solid in a beaker or test tube. The key factor is the solubility of the product in the solvent used. For most microchemical work, water or alcohol typically functions as the solvent, but other compounds can be used. The solids formed may consist of covalent molecular compounds such as solid aspirin or sugar. They can also be ionic solids that form crystal lattices (e.g., salts) of cations and anions. Many solids take more than one stable form and are considered **polymorphic**. Solid polymorphism has attracted great interest in areas such as pharmaceuticals because crystal forms may play major roles in determining efficacy and manufacturing methods. We will discuss polymorphism in detail in Section 9.4.

One of the most common microchemical tests is to add a reagent to a sample and look for the formation of characteristic salts. Because the formation of ionic solids is such an important part of microchemistry, it is worthwhile to review the related concepts, starting with precipitates that form in aqueous solution. The **solu-bility rules** (Table 9.1) include the observations that all nitrate (NO_3^-) and ammo-nium (NH_4^+) compounds and other Group 1 salts are soluble with sodium, lithium, and other cations. Accordingly, many precipitating reagents consist of aqueous (aq) solutions of these ions. For example, a common reagent for detection of the chloride anion (Cl^-) is a solution of silver nitrate [$AgNO_3$(aq)]. Since all nitrates are soluble, no solid forms in the reagent, but when the nitrate is added to a sample containing chloride, AgCl solid(s) forms immediately.

TABLE 9.1
Solubility Rules

Soluble Compounds	Exceptions
Ammonium salts (NH_4^+)	None
Lithium salts	None
Sodium salts	None
Potassium salts	None
Nitrates (NO_3^-)	None
Perchlorates (ClO_4^-)	None
Acetates ($CH_3CO_2^-$)	None
Chlorides, bromides, iodides	Ag^+, Hg_2^{2+}, Pb^{2+} compounds
Sulfates (SO_4^{2-})	Ba^{2+}, Hg_2^{2+}, Pb^{2+} compounds
Insoluble Compounds	
Carbonates (CO_3^{2-})	NH_4^+, Li^+, Na^+, and K^+ compounds
Phosphates (PO_4^{3-})	NH_4^+, Li^+, Na^+, and K^+ compounds
Sulfides	NH_4^+, Li^+, Na^+, and K^+ compounds
Hydroxides (OH^-)	Ba^{2+}, NH_4^+, Li^+, Na^+, and K^+ compounds

Consider what causes a solid to form. A large drop of solution containing silver nitrate ($AgNO_3$) is placed on a microscope slide (Figure 9.1). To this same drop, a drop of a solution containing aqueous NaCl is added. According to our solubility rules, a precipitate (AgCl) should form, but certain conditions must be met first. These conditions can be generalized to all precipitation reactions.

First, AgCl will precipitate out, as long as the concentrations of the ions are sufficiently high. These concentrations are defined by the solubility product constant or K_{sp}:

$$AgCl(s) \rightleftarrows Ag^+ + Cl^-$$

$$K_{sp} = [Ag^+][Cl^-] = 1.8 \times 10^{-10}$$

where the value in brackets represents the molar concentration of each ion (moles/L). K_{sp} describes the equilibrium established between a solution and the solid with which it is in contact. The tiny value of the solubility product constant indicates that AgCl is relatively insoluble and little of it redissolves after it is formed. Whenever

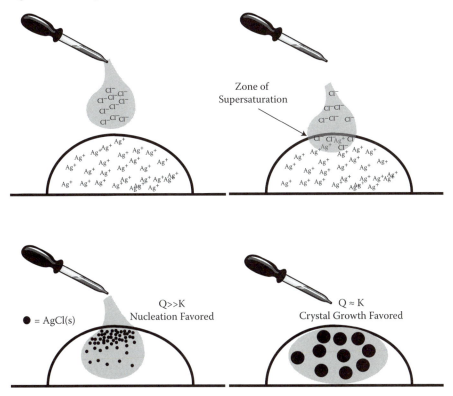

FIGURE 9.1 Formation of solid AgCl from aqueous solutions of $AgNO_3$ and NaCl. Assuming no mixing, when the drops first combine, a zone of high concentration favoring supersaturation and nucleation may occur. Larger crystals are more likely when Q is just slightly larger than K and the mixture is homogeneous.

the concentrations of ions exceed the equilibrium values, the system will seek to re-establish equilibrium through formation of precipitation. Assume that the initial concentrations of silver and chloride are 0.001M or 1×10^{-3}M. When one drop of each is mixed, the volume of the solution approximately doubles and at the frozen moment of mixing, the concentrations of the two ions are now halved (0.0005M each) and no equilibrium is established. Under non-equilibrium conditions (symbolized by **Q**), we can calculate:

$$Q = [Ag^+] [Cl^-] = 0.0005 * 0.0005 = 2.5 \times 10^{-7}$$

By comparing the value of Q to K, we can see how far the system is out of equilibrium. If $Q > K_{sp}$, the system will be driven to restore equilibrium by precipitation, and the process of forming a solid will continue until Q = K. If $Q < K_{sp}$, nothing occurs because the concentrations of one or both ions are too low for a solid to form.

To continue with this example, at the instant of mixing, the solution created by combining the two drops is considered **supersaturated** as shown in Figure 9.1. The system is not stable under these conditions and a solid immediately begins to form.* It will continue to form until equilibrium concentrations are restored. The speed of formation of the particles and particle size depends on how much the equilibrium values are exceeded. This is of importance to microchemistry because of the importance of crystal morphology.

In general, the greater the relative supersaturation, the faster particles will form and the smaller they will be (Figure 9.1, lower left). Crystals typically grow in two stages: (1) **nucleation**, a few ions come together to form a solid; they are obviously very small and will not necessarily settle immediately or at all; (2) **crystal** or **particle growth** can occur when other cations and anions join the nucleus to form everlarger particles. When the relative supersaturation is high (Q >> K), nucleation is strongly favored over particle growth. High concentrations tend to produce lots of very small crystals instead of a few large ones. Larger crystals are usually preferred for microscopic work.

Certain techniques can encourage slower crystal growth and discourage nucleation. First, use solutions that are as dilute as possible. Second, use the largest volume feasible (in microchemistry, volumes are small, usually microliters [µL]). Third, when reagents are combined, unless they are mixed, the solutions will contain concentration gradients and thus areas that are supersaturated. To control this, we can determine how and where mixing occurs. In general, well mixed solutions with Q ~ K favor crystal growth over nucleation, as shown in Figure 9.1. Finally, by heating and cooling a slide, the crystallization rate can be manipulated and the crystals formed can be purified using recrystallization methods analogous to bench-scale work. Three kinds of crystal impurities may be addressed by using fairly simple methods:

---

* Recall that under non-equilibrium conditions, $\Delta G \neq$ zero. In most reaction systems of interest here, ΔG is negative and the reaction will proceed spontaneously as written until equilibrium is restored; Ag^+ (aq) + Cl^-(aq) → AgCl(s).

1. **Surface adsorption** — Ions cling to the crystal surface (but do not bind to it). They can be removed by gentle rinsing with cold water or other cold solvent.

2. **Mixed crystal formation** — One ion in a crystal is replaced by another. For example, if we were making AgCl as in the above example, and bromide was present, it could replace chloride in the crystal because it carries the same charge. This problem is most pronounced when crystals form rapidly or interfering ions are present, which is likely when real samples are studied. Gentle heating of the solid with stirring can sometimes remove the offending ion. This gentle heating is called **digestion**. The idea is that by selecting a solid with a small K_{sp}, the solid will dissolve to some extent in the hot water and slowly reform. However, the offending ion will also dissolve and thus be removed from the crystal. If we force slow crystal formation, the likelihood that the offending ion will be trapped again is much lower.

3. **Occlusion** — Crystals grow rapidly and literally encircle and trap counter ions, dust, solvent droplets, and other materials. These entrapments are often easy to see under a microscope. Clean slides are essential to reduce entrapment to manageable levels. Additionally, digestion that allows slow dissolution and reformation will free trapped materials and hopefully not reintegrate them.

It is worth noting that any molecule with an **ionizable center** can form salts. An ionizable center has the ability to gain or lose a proton to become charged. Thus, the definitions of cation and anion extend well beyond single element species such as Cl^- or Na^+ and simple polyatomics such as NH_4^+ and NO_3^-. Amino acids represent one group of compounds that have ionizable centers; drugs are another. In addition, a molecule may have more than one ionizable center.

To form a cation, a molecule must have an RNH_2 amine group that can be protonated to form RNH_3^+, as shown in the lower frame of Figure 9.2. These are basic compounds such as the alkaloids, a family that includes drugs such as cocaine and methamphetamine. Conversely, acidic molecules that contain acidic protons such as a phenolic OH^- or a carboxylic acid group ($COOH^-$) may, under proper pH conditions, be deprotonated to form an anion with an O^- site. Amino acids are defined by the presence of both groups on the same molecule; other molecules such as morphine may be amphoteric (Figure 9.2, top right).

While it is beyond the scope of this book to discuss the detailed acid and base chemistry of these compounds, it is important to realize that for such molecules to form characteristic salts, the ionizable center must be ionized. This is a function of pH. To simplify this discussion, we will adopt a standard used in the pharmaceutical industry by which all drugs, whether acidic or basic, are described in terms of how easily their ionizable sites lose protons:

$$\text{Acidic drugs: } ROH \leftrightarrow RO^- + H^+ \tag{9-1}$$

$$\text{Basic drugs: } RNH_3^+ \leftrightarrow RNH_2 + H^+ \tag{9-2}$$

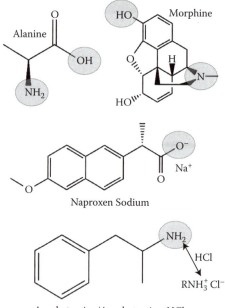

FIGURE 9.2 Molecules with ionizable centers that can form salts. The top two compounds are amphoteric (have acidic and basic sites). Naproxen is an acidic compound and amphetamine is a basic compound.

The only reason we choose to write the equation this way is that it allows us to use the pK_a value as a baseline for determining the pH needed to ensure that an ionizable center is ionized and thus available to form a salt. For basic drugs, the RNH_2 form can be referred to as the free base; the salt is denoted as $RNH_3^+Cl^-$. An example is cocaine that exists as the oily free base form or as cocaine•HCl. Similarly, acidic compounds are often in the form of a Group I salt such as naproxen sodium.

Recall that the K_a value is the acid dissociation constant that is analogous to the solubility product constant (K_{sp}) just described, and the pK_a is the $-\log K_a$. It is calculated for the above equations as:

$$\text{Acidic drugs: } K_a = \frac{[RO^-][H^+]}{[ROH]} \tag{9-3}$$

$$\text{Basic drugs: } K_a = \frac{[RNH_2][H^+]}{[RNH_3]} \tag{9-4}$$

By using the log and rearranging the scheme, we obtain the Henderson–Hasslebach equation that you may recall from a discussion of buffers and acid/base indicators. This form is useful for determining the pH needed to ensure ionization:

$$\text{Acidic drugs: } pK_a - pH = \log \frac{[\text{un – ionized}]}{[\text{ionized}]} \tag{9-5}$$

$$\text{Basic drugs: } pK_a - pH = \log \frac{[\text{ionized}]}{[\text{un-ionized}]} \tag{9-6}$$

If a drug is basic (has an amine group), the pH should be at least two units more acidic than the pK_a. This makes sense; if excess protons are present, the dissociation equilibrium (Equation 9-2) will be driven to the left (ionized) form. Conversely, to ionize acidic drugs, the pH should be two units more basic than the pK_a. In the presence of excess base (OH^-), the protons formed (Equation 9-1) will tend to form water, driving the system to the right (ionized) form.

Why two pH units? The concept corresponds to an excess concentration factor of 100 on a log scale. If two units are good, would more than two units be better? Not necessarily. Extreme pH values (very low or very high) can produce unwanted effects, reactions, and possible degradation of the analytes of interest. It is good practice to have a reasonable estimate of the pH needed for a test and keep the reaction in that range, avoiding pH values of 0 to 2 or 12 to 14 if possible.

Finally, what about compounds like amino acids, drugs, and proteins that have multiple ionizable centers? For microcrystal testing, these compounds can be problematic and a case-by-case approach is best. Morphine (Figure 9.2) has two ionizable centers, one cationic and one anionic. It forms salts, the most common of which is morphine sulfate. From the name, we can deduce that the morphine is the cation and that for charge balance it must carry a charge of 2^+.

9.3 MICROCRYSTAL METHODS

Microchemical tests can be viewed and categorized in several ways. Most generate solids that may be metallic, crystalline, or molecular. For example, many of the common tests used for organic functional group analysis can be modified for microscopic application. Solubility studies can be conducted microscopically, and acid/base indicators can be used to estimate pH. However, in general, microchemical testing involves formation of a precipitate and we will focus on that. We already discussed the theory of solid formation and the importance of pH when ionizable centers are involved. With this information as background, the outline of a microchemical test is simple: apply a solid or liquid sample to a slide; add a solid or liquid reagent; and observe the solid. The specificity of any test depends on the morphology of the crystal, its geometry, one or more refractive indices (RIs), and behavior under polarized light conditions. We will discuss each factor, beginning with sample preparation.

There are many ways to combine the material to be tested and the test reagent on a microscope slide. Table 9.2 summarizes the more common methods. No set rules specify which method is best, although some generalizations can be made.

Ideally, slow crystal growth is desired to yield the best crystals for microscopic study. We already discussed how to encourage slow growth. Knowing the K_{sp} or

TABLE 9.2

Selected Methods for Microchemical Testing

A	Place a drop of a solution containing unknown on a slide and add a drop of test reagent
B	Place a drop of unknown solution on a slide and add a small crystal of solid test reagent
C	Reverse the procedure: to a small solid sample of unknown, add a drop of test reagent
D	Place a drop of unknown solution on a slide and allow it to evaporate (with or without gentle heating); add a drop of test reagent to the residue; the drop-and-heat procedure can be repeated to increase concentration of unknown
D	Place a drop of unknown close to a separate drop of test reagent; using a platinum needle or other tool, draw an edge of one drop into the other
E	Use a capillary tube; allow one solution to be drawn up at one end and the other solution at the other end; reaction will occur in the middle, similar to D above; see Section 9.5 for an example

at least the relative solubility of the expected precipitate represents a good starting place. The AgCl solid is very insoluble and forms so rapidly that barring further treatment, it is difficult to identify any crystal morphology. This solid represents an excellent example of an extreme case. If you were to use this test without further treatment, dilute solutions are needed along with patience to allow the solutions to evaporate sufficiently such that Q > K. Thus, the method of adding a drop to a drop is reasonable a place to start. However, if a solid is relatively soluble or does not form quickly, method D would be the place to start because by successive deposition and evaporation, a significant concentration of solid on the slide could be built up.

A different method can be used with volatile materials such as methamphetamine. Like amphetamine, methamphetamine (Figure 9.2) is a basic compound with a single ionizable amine group. It is stable (oily solid) in the free base form or as a salt, typically a hydrochloride. Several microcrystal tests can be performed on methamphetamine and because the free base has a high vapor pressure, it is possible to create a simple separation and purification step as part of the microcrystal procedure. The raw sample is placed inside a well or ring to which a small amount of base is added. A drop of reagent is placed on a cover slip that is then inverted over the well or ring. As per Equation 9-2, equilibrium favors the volatile free base form. This compound evaporates appreciably at room temperature and the vapors enter the hanging drop of reagent. The cover slip is then removed, gently inverted on another slide, and the crystals observed.

Data about reagents that have been or could be used for microcrystal tests fill several books. Because this chapter is meant as an overview, we will only discuss a few selected tests to illustrate common methods and typical results. If you wish to master chemical microscopy, the classic reference remains Chamot and Mason's *Handbook of Chemical Microscopy, Second Edition*, published in 1939. As of this writing, you can order two complete chemical reagents sets that cover all the experiments in this book from McCrone Microscopes and Accessories (www.mccronemicroscopes.com). These kits (Figure 9.3), supplemented with a few solvents, buffers, and simple tools are ideal for detailed explorations of chemical microscopy.

One discipline that uses microcrystal tests extensively is forensic drug analysis. Microcrystal testing can be a powerful screening test in the hands of a skilled microscopist (Figures 9.4 and 9.5). The international Scientific Working Group for the Analysis of Seized Drugs (SWGDRUG, www.swgdrug.org) classifies microcrystal tests as Category B methods that can be used to confirm other analytical findings. Microcrystalline tests are used mostly for basic drugs in which the cation is the RNH_3^+ species described above, but tests for acidic drugs exist as well. The American Society for Testing and Materials (ASTM) published standards that describe several of these tests (2003, 2006, 2007).

In drug analysis (pharmaceutical and forensic), and organic analysis in general, microcrystal tests can be useful tools for distinguishing optical isomers, determining whether a pure enantiomer or a racemic mixture is present in a sample, and evaluating the results of organic syntheses. As an example, 5-nitrobarbituric acid (also known as dilituric acid) has been used to differentiate a group of phenylalkylamines and phenylalkylamino acids based on crystal morphology, thermal analysis, infrared, and x-ray data (Britttain and Rehman 2005). Racemic mixtures of some of the phenylethylamine drugs (methamphetamine, amphetamine, and related compounds) can sometimes be distinguished from optically pure compositions using reagents containing $HAuCl_4$, $HAuBr_4$, H_2PtBr_6, or H_3BiI_6 (Fulton 1969).

One of the most interesting uses of microchemical testing is through electrochemical (oxidation and reduction) reactions. The reaction results in formation of a metallic solid such as silver or copper, not a salt or crystal. The mechanism of reaction is electron exchange dictated by the relative reduction potentials of the metals of interest (Table 9.3). Elements that are the best competitors for electrons and thus the most likely to be in the metallic state and least likely to surrender

FIGURE 9.3 Chemical microscopy kits. The few milligrams in each reagent bottle will last for years.

FIGURE 9.4 Crystals formed by gold chloride and cocaine, here created in an agarose gel. This is a useful technique for obtaining images showing crystal growth habits.

FIGURE 9.5 Crossed polar view of crystals formed by γ-valerolactone (GHV, a GHB analog), and $AgNO_3$. Note how thin these crystals are.

electrons appear at the bottom of the table. This makes sense; gold, platinum, and palladium are extraordinarily stable and relatively inert. The elements most likely to surrender electrons and be in the ionic state are listed at the top of the table. Using the table is simple. Consider, for example, a drop on a slide containing aqueous Ag^+. If a tiny piece of zinc metal is dropped into this solution (acidified with HCl), you would observe the zinc dissolving and branches of solid silver forming. The silver is the stronger competitor for electrons and takes them away from the zinc metal; the silver is reduced and the zinc is oxidized. This process is shown as a time series of images presented in Figure 9.6. Silver is growing outward from a piece of zinc; the photos were taken about 20 seconds apart.

Microcrystals, whether ionic lattices or molecular compounds, can be formed and evaluated after recrystallization—the simple process of dissolving and reforming a solid under controlled conditions (Figure 9.7). These conditions are usually designed to form large, stable, purified crystals with minimal inclusions and occlusions. Any reagent in which the solid is soluble can be used as a recrystallization solvent, including water, alcohols, organic solvents, and even acids.

As an example and a way to conclude this section, let us return to silver chloride—a compound so insoluble that even under carefully controlled conditions nucleation often leads to a milky cloud of indistinct particle morphology.

TABLE 9.3

Activity Series for Electrochemical Testing

Metal	Approximate Half-Cell Potential[a]	
Ca	− 2.87	Least likely to be reduced
Mg	− 2.36	Least electronegative
Al	− 1.68	
Cr	− 0.74	
Mn	− 1.18	
Zn	− 0.76	
Cd	− 0.40	
Fe	− 0.44	
Co	− 0.28	
Ni	− 0.24	
Sn	− 0.14	
Pb	− 0.13	
H_2	0.00	
Cu	0.34	
Hg	0.85	
Ag	0.80	
Pd	0.92	
Pt	1.18	Most electronegative
Au	1.69	Most likely to be reduced (metal)

[a] Standard conditions.

FIGURE 9.6 Electrochemical reaction of aqueous silver and solid copper.

Recrystallization from a solution of ammonium hydroxide is an effective way to create better formed crystals by slowing the reaction process. When NH_4OH is added to the AgCl, a strong complex forms. The ammonia then slowly evaporates, allowing the slow regrowth of larger silver chloride crystals that are amenable to microchemical identification.

9.4 POLYMORPHS, HYDRATES, AND CO-CRYSTALS

One of the most useful applications of microchemical methods is in the study of polymorphism of crystals. A polymorphic crystal (salt, molecular compound, or element) has more than one stable crystal form or habit (Stahly 2007). As described

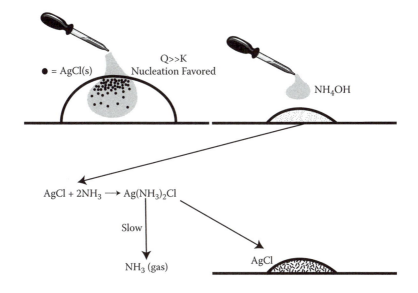

FIGURE 9.7 Silver chloride recrystallization.

above, crystallization begins with nucleation in which a small number of molecules or cation–anion combinations aggregate in a pattern similar to that of the larger stable lattice (Rodriguez-Hornedo and Murphy 1999; Rodriguez-Spong et al. 2004). The larger lattice then grows (crystal growth stage described above) on this superstructure. The more supersaturated the system, the more nucleation is favored.

The key factors controlling the process are (1) the solubility of the crystal in any solvents present, (2) the degree of supersaturation, (3) the ability of ions or molecules to move toward the nucleus, (4) the temperature, and (5) the reactivity of any surfaces present toward nucleation (Rodriguez-Hornedo and Murphy 1999). This is a simplified description of a complicated process. If you are interested in a more detailed theoretical treatment of this topic, the cited articles (Price 2008; Rodriguez-Hornedo and Murphy 1999; Rodriguez-Spong et al. 2004) are excellent resources.

Polymorphs are not the same as **co-crystals** that contain different atoms, different ions, or different molecules (Stahly 2007). The **pseudopolymorph** term includes hydrates and solvated solid forms (Stahly 2007). For example, magnesium sulfate is a stable solid in the anhydrous $MgSO_4$ form. Epsom salts ($MgSO_4*7H_2O$) have a heptahydrate form. Chemically, the cation–anion combination is the same but each cation–anion pair has seven water molecules stoichiometrically associated with it. In many cases, hydrate forms can be converted to anhydrous forms by heat. However, under a microscope, the crystal forms may be very different even if the compound of interest is the same. Solvents may also be stoichiometrically associated with a salt in an analogous way; this is not the same as the solvent occlusions or inclusions described earlier.

Polymorphism of pharmaceuticals is currently an area of great interest but the phenomenon was noted over 200 years ago in relation to elemental forms called allotropes (Stahly 2007). Graphite and diamond are allotropes of carbon, for example. In the case of a drug, the polymorph form can have a dramatic effect on pharmacological activity. Many explosives are also polymorphic. Fundamentally, polymorphism arises when atoms and molecules can assume more than one stable arrangement as a function of inter- and intra-molecular forces at work within the solid. The polymorph formed is often determined by the crystallization or recrystallization conditions or by temperature.

A familiar drug with stable polymorphic forms is acetaminophen (paracetamol or Tylenol®) (Barthe, Grover, and Rousseau 2008; Nichols and Frampton 1998). Three forms have been reported; two are fairly easy to crystallize with careful control of experimental conditions. Form I is monoclinic and form II is orthorhombic (Figure 9.8). Form I is found in Tylenol tablets (Nichols and Frampton 1998). With careful control of solvents and cooling rates, transitions between the two polymorphs can be directly observed and quantified using image analysis. The actual structure of the crystal is determined using x-ray diffraction as shown in Figure 9.9. The differences are slight, but as shown in Figure 9.10, even slight differences can lead to significantly different crystal morphologies.

It is possible to track the crystallization process as shown in Figure 9.11. Using reflectance and image analysis, the authors were able to track different stages in the crystal formation and reformation processes. In the top frame, the particle counts and sizes rise abruptly, corresponding to the onset of nucleation. At about 50 minutes (first photomicrograph), lots of small needle-shaped form II crystals are noted (see scale on figure). At point (b), both form I and form II crystals are

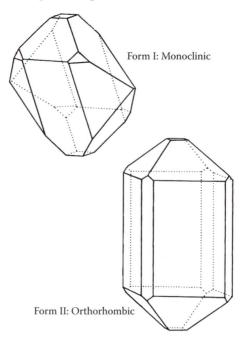

Form I: Monoclinic

Form II: Orthorhombic

FIGURE 9.8   Two crystal forms of paracetamol.

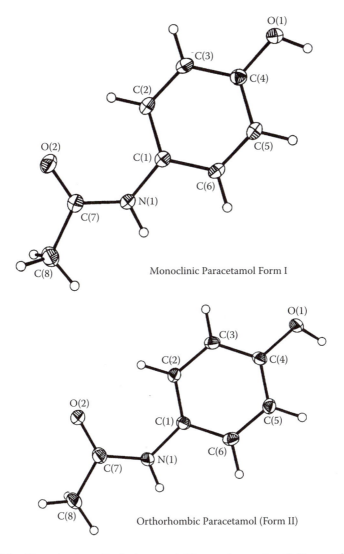

Monoclinic Paracetamol Form I

Orthorhombic Paracetamol (Form II)

FIGURE 9.9 X-ray structures for the two crystal forms of paracetamol. (*Source:* G. Nichols and C.S. Frampton, *Journal of Pharmaceutical Sciences,* 87, 684–693, 1998. Reproduced with permission of American Chemical Society and American Pharmaceutical Association.)

seen. By point (c), a reasonably steady state has been established and form II crystals are present.

Another polymorphic substance is nabumetone, a member of the nonsteroidal anti-inflammatory drug (NSAID) family. There are two common crystal forms that, like acetaminophen, exhibit distinctly different morphologies. The solid forms have different properties. These differences arise solely from the way the molecules are arranged in the crystal lattice. Figure 9.12 depicts the microscopic appearances of forms I and II; Figure 9.13 shows the corresponding crystal packing diagrams. The

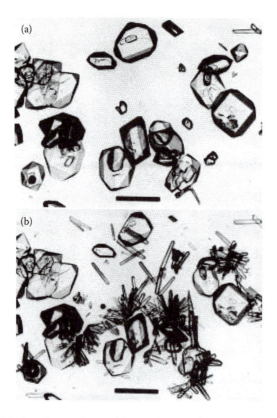

FIGURE 9.10 Solution phase polymorphism and conversion of orthorhombic paracetamol (needles) to monoclinic prisms and plates. Images were taken at room temperature, 30 minutes apart. The scale bar corresponds to 500 μm. (*Source:* G. Nichols and C.S. Frampton, *Journal of Pharmaceutical Sciences,* 87, 684–693, 1998. Reproduced with permission of American Chemical Society and American Pharmaceutical Association.)

same form II plates can also be created using capillary tube crystallization, as shown in Figure 9.14. Capillary crystallization has also been used for proteins (Chen, Gerdts, and Ismagilov 2005; Ng, Gavira, and Garcia-Ruiz 2003; Sauter, Dhouib, and Lorber 2007; Warkentin et al. 2008; Yadav et al. 2005; Zheng, Gerdts, and Ismagilov 2005), drugs, and general microchemical analysis. Another method of observing transitions between polymorphic forms is cooling, as shown in Figure 9.15. The transition of form I to form II, from frame (a) to frame (d), took about 1 minute.

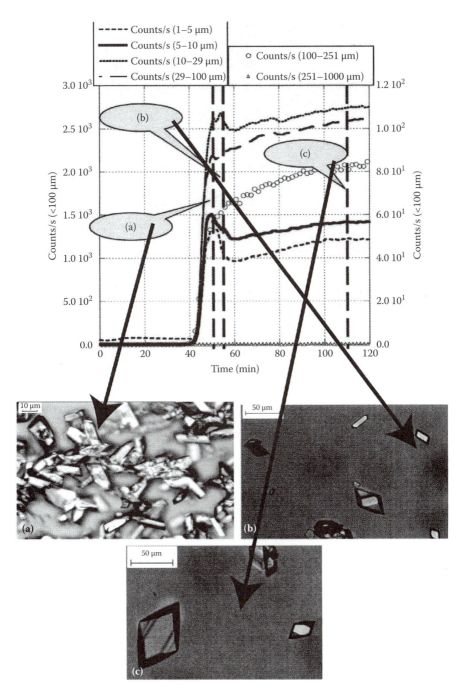

FIGURE 9.11 Paracetamol crystal formation over time. (*Source:* S.C. Barthe, M.A. Grover, and R.W. Rousseau, *Crystal Growth and Design*, 8, 3316–3322, 2008. Reproduced with permission of American Chemical Society.)

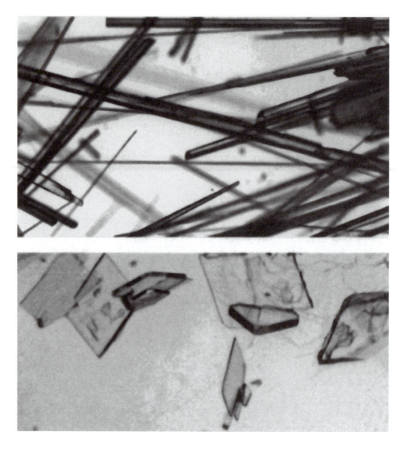

FIGURE 9.12 Two crystal forms of nabumetone. (*Source:* C.P. Price et al., *Crystal Growth and Design,* 2, 501–503, 2002. Reproduced with permission of American Chemical Society.)

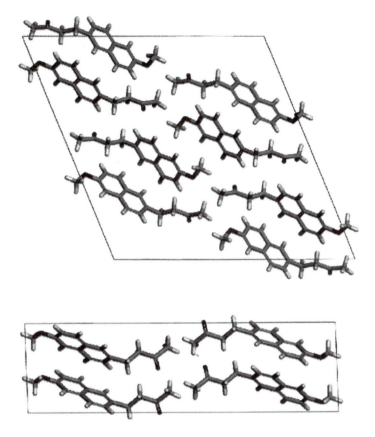

FIGURE 9.13 Crystal packing diagram for two crystal forms of nabumetone. (*Source:* C.P. Price et al., *Crystal Growth and Design,* 2, 501–503, 2002. Reproduced with permission of American Chemical Society.)

FIGURE 9.14   Formation of form II nabumetone crystals in a capillary tube. (*Source:* L.J. Chyall et al., *Crystal Growth and Design,* 2, 505–510, 2002. Reproduced with permission of American Chemical Society.)

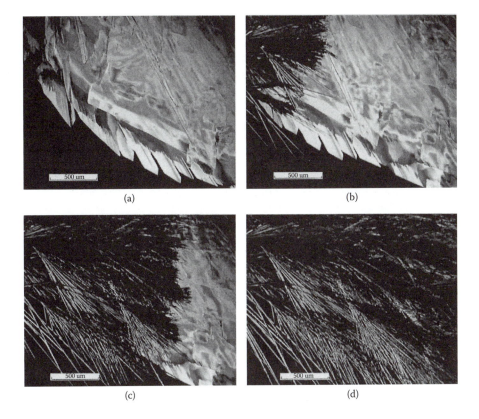

(a)

(b)

(c)

(d)

FIGURE 9.15 Transformation during cooling. (*Source:* L.J. Chyall et al., *Crystal Growth and Design,* 2, 505–510, 2002. Reproduced with permission of American Chemical Society.)

REFERENCES AND FURTHER READING

ASTM, *Standard guide for microcrystal testing in the forensic analysis of cocaine*, 2003.

ASTM, *Standard guide for microcrystal testing in the forensic analysis of methamphetamine and amphetamine*, 2006.

ASTM, *Standard guide for microcrystal testing in the forensic analysis of phencyclidine and its analogues*, 2007.

S.C. Barthe, M.A. Grover, and R.W. Rousseau, Observation of polymorphic change through analysis of FBRM data: Transformation of paracetamol from form II to form I, *Crystal Growth and Design*, 8: 3316–3322, 2008.

H.G. Britttain and M. Rehman, Foundations of chemical microscopy 3: Derivatives of some chiral phenylalkylamines and phenylalkylamino acids with 5-nitrobarbituric acid, *Chirality* 17: 89–98, 2005.

E.M. Chamot and W.M. Mason, *Handbook of Chemical Microscopy*, Vol. 1, McCrone Research Institute, 1931.

D.L. Chen, C.J. Gerdts, and R.F. Ismagilov, Using microfluidics to observe the effect of mixing on nucleation of protein crystals, *Journal of the American Chemical Society* 127: 9672–9673, 2005.

C.C. Fulton, *Modern microcrystal tests for drugs*, Wiley Interscience, 1969.

J.D. Ng, J.A. Gavira, and J.M. Garcia-Ruiz, Protein crystallization by capillary counterdiffusion for applied crystallographic structure determination, *Journal of Structural Biology* 142: 218–231, 2003.

G. Nichols and C.S. Frampton, Physicochemical characterization of the orthorhombic polymorph of paracetamol crystallized from solution, *Journal of Pharmaceutical Sciences* 87: 684–693, 1998.

S.L. Price, 2008. Computational prediction of organic crystal structures and polymorphism, *International Reviews in Physical Chemistry* 27: 541–568, 2008.

N. Rodriguez-Hornedo and D. Murphy, Significance of controlling crystallization mechanisms and kinetics in pharmaceutical systems, *Journal of Pharmaceutical Sciences* 88: 651–660, 1999.

B. Rodriguez-Spong et al. General principles of pharmaceutical solid polymorphism: A supramolecular perspective, *Advanced Drug Delivery Reviews* 56: 241–274, 2004.

C. Sauter, K. Dhouib, and B. Lorber, From macrofluidics to microfluidics for the crystallization of biological macromolecules, *Crystal Growth and Design* 7: 2247–2250, 2007.

G.P. Stahly, Diversity in single- and multiple-component crystals: The search for and prevalence of polymorphs and co-crystals. *Crystal Growth and Design* 7: 1007–1026, 2007.

M. Warkentin, V. Stanislavskaia, K. Hammes, and R.E. Thorne, Cryocrystallography in capillaries: Critical glycerol concentrations and cooling rates. *Journal of Applied Crystallography* 41: 791–797, 2008.

M.K. Yadav et al., In situ data collection and structure refinement from microcapillary protein crystallization. *Journal of Applied Crystallography* 38: 900–905, 2005.

B. Zheng, C.J. Gerdts, and R.F. Ismagilov, Using nanoliter plugs in microfluidics to facilitate and understand protein crystallization. *Current Opinions in Structural Biology* 15: 548–555, 2005.

QUESTIONS

1. In the pharmaceutical industry, it is common for new drugs to be screened for polymorphic forms by melting a pure solid and then observing the crystal forms. No solvents are used. Why?

2. The solids listed in the table below may be present in an unknown white powder. Develop a simple precipitation (or other observable) reaction and describe the method (per Table 9.2) and reagents you would use. Justify your selections.

Solid	K_{sp}
$CaCO_3$	4.5×10^{-9}
$CaSO_4$	2.4×10^{-5}

3. What are some practical limitations of microchemical analysis for investigating unknowns such as white powders?

EXERCISES

Exercise 9.1: Microsolubility

Materials needed:

Naproxen sodium, pK_a = 4.15
Caffeine, pK_a = 10.4
Lidocaine (hydrochloride salt), pK_a = 7.9
Aspirin, pK_a = 3.49
Other compounds as desired
Dropper bottles of water, acetone, methanol, and hexane
Dropper bottles of pH-adjusted water at approximately 2, 4, 6, 8, and 10*

Stock solutions:

Solution	Quantity
0.1M potassium hydrogen phthalate	250 mL
0.1M potassium dihydrogen phthalate	250 mL
0.255M borax	250 mL
0.1M NaOH	100 mL
0.1M HCl	100 mL
0.2M HCl	50 mL
0.2M KCl	50 mL

Buffer solutions:

pH	Solution
2.0	25 mL 0.2M KCl + 6.5 mL 0.2M HCl
4.0	50 mL 0.1M potassium hydrogen phthalate + 1.0 mL 0.1M NaOH
6.0	50 mL 0.1M potassium hydrogen phthalate + 44 mL 0.1M NaOH
8.0	50.0 mL 0.025M borax + 20.5 mL 0.1M HCl
10.0	50.0 mL 0.025M borax + 18.3 mL 0.1M NaOH

Before beginning this exercise, look up the solubilities of the white powder materials. Note that the salt form and free base may have different solubilities. Using this information and Equations 9-5 and 9-6, you should be able to predict what will occur when you test solubilities. First, test each powder in water and the organic solvents. Place a small crystal on a slide and a drop of the solvent a few millimeters away. Draw the crystal into the solvent and note your observations. Repeat with the buffer solutions. Do your results agree with predictions?

Exercise 9.2: Electrochemical series

Materials needed:

0.1M solutions of various ions such as Zn^{2+}, Cu^{2+}, and Ag^+ (prepared if possible from nitrate salts)
Metals (Cu, Zn, Pb, Mn, and Fe) in form of wools, shavings, cuttings, or very small pieces
Dropper bottle of 0.1M HCl

---

* Recommended procedures adapted from *CRC Handbook of Chemistry and Physics*, 84th ed. 2003–2004, Section 8.45.

Using Table 9.3 and the solutions and metals at your disposal, devise mixtures that should generate formation of a new metal by dissolution of another. To perform each experiment, add a drop of HCl to a drop of test solution and then draw the solution to cover the metal. Observe what happens.

Exercise 9.3: Silver chloride

Materials Needed:

Dropper bottles of water, 0.1M $AgNO_3$, NaCl, and concentrated NH_4OH
Well slide or glass ring to create a small well on a slide

Combine one drop of the AgCl and NaCl solutions and observe what happens. Does the appearance of the precipitate change over time (5 to 10 minutes)? Try diluting the reagents on the slide by first placing a drop of the silver solution and a drop of water at one end of the slide. Mix. Repeat with the NaCl solution and water on the other end of the slide. Try mixing these solutions. Vary the water amounts as feasible and note your observations.

Mix one drop of each AgCl and NaCl on a fresh slide. Add one drop of concentrated NH_4OH and note your observations. Allow this slide to sit up to 30 minutes and note your observations. Finally, repeat the experiment on a well slide or inside the glass ring, but cover the solution immediately after adding the NH_4OH. What is the purpose of doing this? Allow this slide to sit for some time, even overnight, and observe the changes. Comment on all the procedures and relate back to your knowledge of crystal growth and formation.

Appendix 1: Answers to Questions

CHAPTER 1

1. When light leaves one medium (such as air) and enters another (such as water), the speed of propagation changes. Why? Explain on the molecular level.

   The speed changes because the electromagnetic environment changes. Because light is an oscillating electromagnetic wave, attractive and repulsive interactions between the wave and matter (composed of atoms) occur constantly. Atoms in turn are composed of nuclei containing protons and electron clouds. Any change in that environment results in a change in the speed of propagation. This speed is maximized in a vacuum because no matter exists to cause such interactions.

2. Based on the concept of diffraction as illustrated in Figure 1.7, what are the implications for designing microscope lenses?

   The size of the lens is very important in determining how much light is emitted and to what degree diffraction occurs. Because diffraction will inevitably occur at any junction, it is important that the lens and the optical train in general be designed to account for diffraction. As seen in Figure 1.7, a fairly small zone of maximum intensity of diffracted light emerges from a small aperture. Lens designs must account for and capture as much of this light as possible.

CHAPTER 2

1. Consider a situation in which light impinges on a paraxial lens at an angle of 35°. Construct a ray diagram indicating the focal point (show x and y coordinates) of this situation. Assume that you are using a positive lens with a focal length of 45mm.

   See Figure A1.1. The source is set at infinity; therefore all incoming rays are parallel. The rays will intersect the optical axis at 35°. Thus the offset of the focal point from the optical axis is $y = f * \tan\theta = 45 \text{ mm} * \tan(35°) = 31.5$ mm. The position of the focal length on the optical axis remains 45 mm.

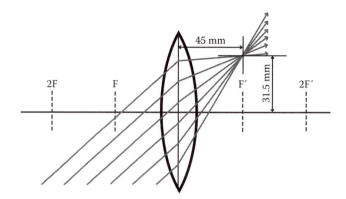

FIGURE A1.1   Ray diagram.

2. For a positive lens with focal lengths of 10, 15, 20, and 30 mm, calculate the values for *b* if the values for *a* are 35, 60, 90, and 120 mm. For each situation, calculate the magnification factor.

From the thin lens equation (Equation 2-2) we can calculate the value of *b* by

$$\frac{1}{f} = \frac{1}{a} + \frac{1}{b}$$

$$\frac{1}{b} = \frac{1}{f} - \frac{1}{a}$$

$$\frac{1}{b} = \frac{(a - f)}{af}$$

$$b = \frac{af}{(a - f)}$$

From Equation 2-3 we know that the magnification is the ratio of *b* to *a*. Values of *b*:

b/mm	*a* = 35 mm	*a* = 60 mm	*a* = 90 mm	*a* = 120 mm
f = 10 mm	14.0	12.0	11.3	10.9
f = 15 mm	26.3	20.0	18.0	17.1
f = 20 mm	46.7	30.0	25.7	24.0
f = 30 mm	210.0	60.0	45.0	40.0

Magnification:

Magnification	a = 35 mm	a = 60 mm	a = 90 mm	a= 120 mm
f = 10 mm	0.4	0.2	0.1	0.1
f = 15 mm	0.8	0.3	0.2	0.1
f = 20 mm	1.3	0.5	0.3	0.2
f = 30 mm	6.0	1.0	0.30	0.3

CHAPTER 3

1. Why is the objective lens the most important optical component of a microscope?

The objective lens is fundamentally the determinant of resolution. The purpose of a microscope is to allow the observation of fine detail, and the observation power is controlled primarily by the objective. Depending on the application, different types of objectives ranging from relatively low correction achromats to highly corrected apochromats may be selected.

2. Which components of a microscope provide the necessary magnification, resolution, and contrast and why?

The magnification of a microscope is the product of the magnification of the objective and the eyepieces. Objectives typically range from 500 to 1000× the *NA* of the objective. Thus an objective with a *NA* of 0.65 will have an optimum total magnification of 325 to 650×. A rule of thumb is that the magnification of the objective must be higher than the magnifications of the eyepieces; the opposite combination achieves the same magnification with reduced resolution. The resolution of a microscope is determined by the *NA* of the objective and the condenser. We know that the resolution of a microscope for a certain wavelength is inversely proportional to the *NA* of the objective. Thus the larger the *NA*, the better the resolution of the system.

If the *NA* of the condenser is less than that of the objective, the gain in *NA* (and thus resolution) is limited by that of the condenser. If optics with *NA* values >1.00 are used dry, the added benefit of the increase in *NA* is lost. Resolution is also affected, although indirectly, by the condenser aperture. Since the condenser aperture controls the cone of light (*NA*) entering the objective, it can reduce the effective *NA* of the system and thus its resolution. The condenser also imparts contrast to the image of the specimen. One of the advantages of the condenser aperture is that, when its size is changed, the increase or decrease in illumination is uniform across the field of view. If there is little inherent contrast in the specimen, contrast can also be increased by choosing a suitable mechanism such as dark field, phase contrast, or polarized light.

3. What role does dispersion play in the quality of an objective lens?

When the refractive index (RI) of a material is dependent upon the wavelength of light passing through it, the material displays dispersion—the cause of chromatic aberrations in lenses. Adding additional lens elements to an objective decreases the effects of dispersion. Apochromatic lenses have more corrections for dispersion than do achromatic lenses.

4. How can a lens be corrected for spherical aberrations?

Spherical aberrations in cameras are reduced by using aspherical lenses that adjust the lens thickness so as adjust the path length through the glass. The process in a microscope is more complex. A single glass lens, even with monochromatic light, can never be free of spherical aberrations. These defects are caused by the shape of the lens. The best method of reducing such aberrations is by making use of a number of lenses of the necessary shapes and radii that act together to give the desired result. The degree of spherical aberration is inversely proportional to the number of lenses in the optical system.

5. Why should the numerical aperture of the condenser be greater than or equal to the numerical aperture of the largest objective?

The size of the cone of light that enters the objective is controlled by the condenser and its diaphragm. Thus, if the *NA* of the condenser is significantly smaller than that of the objective, the cone of light produced by the condenser is insufficient to cover the entire front lens of the objective. This will reduce the effective *NA* of the objective and it will perform below specification.

CHAPTER 4

1. When using a filter near a field diaphragm, it is very important that the filter be very clean. Why?

The field diaphragm is a conjugate plane in the image ray set in Kohler Illumination. We focus the field diaphragm by adjusting the condenser position. Since the field diaphragm is in focus in the image plane, anything near it will be slightly defocused in the image plane. Any dust particles on a filter near a field diaphragm will appear as defocused spots in the image.

2. The collector lens on the lamp holder can be used to focus the filament. Why do we want to do this?

When setting up Kohler Illumination, we focus the filament by inserting the Bertrand lens and adjusting the collector lens. We also center the position of the filament that forms portions of the illumination rays and conjugate planes. The focus adjustment ensures that the filament is not focused in the plane of the specimen and provides uniform illumination to the specimen.

3. We can view the conjugate planes of the illuminating rays by inserting a Bertrand lens. This can also be achieved by removing an eyepiece. Why does the eyepiece removal also work?

When the eyepiece is removed, the microscopist can focus on the back focal plane of the objective which is conjugate with the aperture diaphragm and the filament. If you open and close the aperture diaphragm you will see the effect. You should also be able to view the filament. Although this can be done directly, it is easier to do so with the Bertrand lens or a focusing telescope.

4. After you have set up Kohler Illumination, you can reduce the brightness of the image by closing the field diaphragm or defocusing the condenser. Is this the correct technique?

No. This would defeat the object of Kohler Illumination. The best way to reduce the brightness would be to insert a neutral density filter into the optical path of the microscope. Reducing the voltage on the lamp is not a good solution because it causes a color cast in the image. Any other action would reduce the quality of the illumination of the sample, introduce artifacts into the image, and destroy resolution.

5. What are the advantages of Kohler Illumination?

The main advantage of Kohler Illumination is bright and even illumination. Its main characteristics are the projection of an image of the light source in the plane of the condenser diaphragm and the projection of an image of the field diaphragm in the plane of the specimen. The focal length of the condenser ensures that the image of the filament fills the aperture diaphragm. The field diaphragm controls the amount of light reaching the sample (remember we extended the diaphragm until it was just outside the field of view). The control of the field diaphragm allows the control of glare in the system. The aperture diaphragm controls the convergence of the light (it can also adjust intensity but must not be used for this purpose). If the image of the filament does not fill the aperture diaphragm, the aperture diaphragm is reduced and resolution is lost. If the filament is not centered, the result in effect is oblique illumination.

6. The numerical aperture of a microscope equals the average of the *NA* of the objective and the condenser. If the *NA* of the condenser is 1.25, when would it be appropriate to use this equation?

Equation 2-9 states that the resolution of an objective is given by

$$d = \frac{0.61\lambda}{NA_{objective}}$$

when the *NA* of the condenser is greater than the *NA* of the objective. In this case the $NA_{objective}$ is not limited by the $NA_{condenser}$. When the $NA_{condenser}$ is less than the $NA_{objective}$, the $NA_{objective}$ is limited. In this case Equation 2-10 states

$$d = \frac{1.22\lambda}{NA_{condensor} + NA_{objective}}$$

If the $NA_{condenser} = 1.25$ (an oil immersion condenser), it would be inappropriate to use the average since the limiting factor will be the $NA_{objective}$. This applies to any situation in which the $NA_{objective} \leq 1.25$. Obviously for objectives with $NA_{objective}$ values ≥ 1.0, you must use immersion oil (for both objective and condenser) to achieve full resolution.

CHAPTER 5

1. Assume you work in a forensic laboratory as a trace evidence analyst. Crime scene technicians deliver a bag from a vacuum cleaner used to collect trace evidence from a car. Describe the sampling challenge, how you would sample, and how you would justify your approach.

Since you did not collect the evidence, you must rely on the information provided by the evidence collection technician. The key is to obtain representative samples from what inevitably will be a complex mixture of hair, fibers, dust, dirt, glass, etc. One approach is to spread a large square of butcher paper on a table and carefully pour the contents of the bag out onto it. The material can be teased apart, spread over the paper, and examined first with a magnifying lamp to identify any unusual or noteworthy larger items. Beyond that, a grid system could be used to divide the material into subareas to be sampled carefully and thoughtfully. Hairs, fibers, dirt, and dust are likely to represent the bulk of the bag contents and thus require special attention to ensure that representative samples of each substance are collected—a difficult sampling challenge!

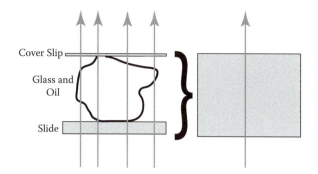

FIGURE A1.2   Absence of refraction due to lack of contrast.

2. Assume you have glass fragments on a glass slide, covered with a glass cover slip mounted in a medium with the same RI value as all the glasses used. Draw a figure to show why you cannot see the glass sample.

See Figure A1.2. If the RIs of the sample, glass, cover slip, and slide are the same, the electromagnetic environment experienced by the light never changes, meaning that no refraction occurs as light moves from one medium to another. As a result, the light may as well pass through a solid block of material with a single uniform refractive index. No delineation of edges will appear; the sample will not be visible because of the lack of contrast between it and the medium.

3. Repeat the above exercise two more times. In the first case, assume the mounting medium has a lower refractive index than the sample. In the second, assume the RI of the medium is higher.

The RI determines the direction in which light will bend, as shown in Figures A1.3 and A1.4. This refractive interaction will increase the contrast between sample edge and medium, making the sample more visible. The greater the relative difference in RIs, the greater the refraction and the sharper the contrast.

CHAPTER 6

1. Why is it difficult to focus an image through a microscope when a high numerical aperture objective is used?

Equation 2-11 states

$$DOF = \frac{n\lambda}{NA^2}$$

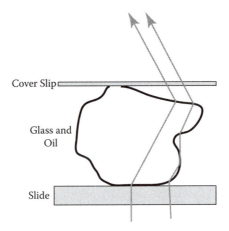

FIGURE A1.3   Effect of refractive index on contrast.

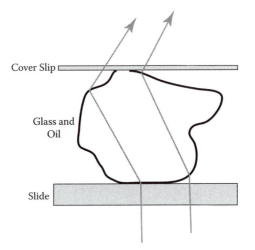

FIGURE A1.4   Effect of refractive index on contrast.

We can see that the depth of field is inversely proportional to the square of the numerical aperture. Table 2.1 gives an indication of the depth of field for a 0.85 (dry) aperture as 0.768 µm. This can be significantly less than the thickness of the specimen. It may be impossible to have the complete specimen in focus at once.

2. If your camera can only record JPG images, why is it important to convert them to a TIFF format as a first step?

The JPG format is a lossy compression, and thus each subsequent save of the image will reduce its quality. Saving as a TIFF (lossless compression algorithm) will prevent loss of information and the image will not degrade over time.

3. Why would you prefer to record an image as 16-bit even if image processing software can work only with 8-bit images?

The area of interest in an image may span only a few grayscales in an 8-bit image. Selection of a specific area in a 16-bit image will allow the available grayscales to fill the dynamic range of an 8-bit image (256 levels).

4. The color balance in an image can be adjusted by the camera. In photomicrography, in what other way can it be adjusted?

Photomicrography color balance can be adjusted in four ways: (1) using the adjustment of the camera, (2) adjusting the color temperature of the lamp (voltage), (3) inserting an appropriate filter into the optical path, and (4) during post processing. The on-board microscope solutions are adjustments of the voltage and the filter. The voltage will assist to some extent in terms of intensity, but since most filaments are made of tungsten, the image will always have a typical tungsten cast. The cast can be removed by inserting a daylight filter.

5. We know that the median of a group of numbers is the middle number when the group is sorted in ascending order. Explain how you think a 3×3 median filter works.

Imagine that the following matrix represents a 3×3 median filter applied to a section of an image:

$$\begin{bmatrix} 12 & 13 & 17 \\ 25 & 27 & 22 \\ 25 & 28 & 30 \end{bmatrix}$$

Each element of the matrix represents a pixel in the image. The pixel of interest of the median filter is the central one (grayscale = 27). We can arrange the grayscale values in increasing order:

$$\{12; 13; 17; 22; \mathbf{25}; 25; 27; 28; 30\}$$

The median value of this list of grayscales is 25. In the processed image, the grayscale value of 27 would be replaced by a value of 25. This median filter would then move on to assess the next pixel of interest.

CHAPTER 7

1. Explain the trend seen in Table 7.1., i.e., why does refractive index change as a function of wavelength?

   As seen in the table, refraction decreases with increasing wavelength. The increase of the apex-to-apex distance between wavelengths changes the way this oscillating electromagnetic wave interacts with the electromagnetic environment of the medium. Longer wavelengths interact less relative to shorter wavelengths.

2. Why is birefringence a wavelength-dependent phenomenon?

   Birefringence is fundamentally a refractive index phenomenon, so the answer to the previous question applies.

3. On a molecular level, explain why a beam of polarized light propagating down the optical axis of a calcite crystal is **not** split.

   It is not split because the crystal structure "appears" symmetrical to the light beam propagating through it. There is no difference in the north–south or east–west direction and therefore no activity to initiate the beam splitting.

4. What would the equivalent image for Figure 7.5 look like for an isotropic crystal such as NaCl?

   The O and E circles would be superimposed on each other and show no discernible difference.

5. When cotton fibers are viewed under crossed polars, some regions show interference colors and some do not. Explain on a molecular level why this is so.

   The fiber is a heterogeneous material; it must have regions that are ordered due to cellulosic polymer structure and areas that are not ordered.

6. Some commercial fibers have cross-sections that look like a dumb bells. How would the cross-sections appear under crossed polars, assuming the fiber is birefringent?

   There would be colored regions on either outside edge, symmetrical on both sides with a zone in the middle showing different interference colors corresponding to the thinner region through which the light passes relative to the thicker edges.

CHAPTER 8

1. Describe the refractive indices in isotropic and anisotropic substances.

   Isotropic substances are defined by a single refractive index (RI). Imagine three sets of axes. If the RI along each axis was equal, its shape would be described as spherical. Anisotropic substances have more than one RI. Uniaxial substances are characterized by two RIs and biaxial substances are characterized by three. Consider a three-axis system for a uniaxial substance; if the value of the refractive index on the z axis was greater than the values on the x and y axes (which are equal), the solid describing this situation would be an ellipsoid. If, in the case of a biaxial substance, the RIs differed, the shape would be a triaxial ellipsoid.

2. Describe how you would determine the refractive index of ω for an optically negative uniaxial crystal.

   A uniaxial crystal would have a tetragonal or hexagonal system. In the case of an optically negative crystal, the value of $\omega > \varepsilon$. If a microscope is set up with plane polarized light and the crystal is oriented on its side (c axis or ε perpendicular to the optical axis of the microscope), you can evaluate the RIs of ω and ε relative to the mounting medium. If the RI of the mounting medium is changed to be equal to either ω or ε, extinction will be reached at 180° intervals for ω or ε, which is equal to the RI of the mounting medium.

   To determine the value of ω we must adjust the RI until extinction is found for all crystals when they are rotated 180°. To determine ε, the RI of the mounting medium is reduced (it is a negative crystal) until no lower RIs are found. It is important to note that some of the crystals may indicate a refractive index (ε') somewhere between ω and ε. The crystal may be tilted on one of its faces (pyramidal or rhombohedral) so that the presentation of ε is reduced. Due to the shape of the tetragonal and hexagonal crystals, the value of ω should always be correct (ω is the same in all directions). To summarize, if we adjust the RI of a mounting medium to a maximum so that all crystals in the field of view exhibit extinction during rotation, the RI is that of ω and the crystal is optically negative. Use a negative uniaxial indicatrix (Figure A1.5) to help you understand this situation.

3. Describe what you would see if the crystal described in Question 2 was mounted in a medium with an RI between ε and ω and the stage was rotated.

   If a uniaxial crystal was rotated, the contrast of the image would vary. The highest contrast would indicate the greatest difference in RIs of the medium and the crystal. This would also indicate the directions of the ω

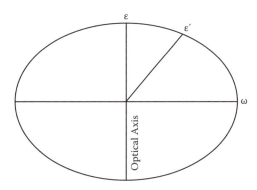

FIGURE A1.5   Negative uniaxial indicatrix.

and ε refractive indices. Evaluating the shape of the crystal would give an indication of the direction of the ω ray. Also, during rotation, the contrast of the ω direction would remain constant, while the contrast in the ε direction would vary based on the orientation of the crystal and indicate whether the crystal was optically positive or negative.

4. If only a few of the crystals of an unknown reach extinction during rotation of the stage, the crystal must be biaxial. Why is this so?

   If the RIs indicated by extinction are those of α (lowest) and γ (highest), the other crystals are not oriented on a plane of the RI and thus they could be α′, γ′, or even β. The assumption here is that the unknown not a mixture of crystals.

5. A triclinic crystal is rotated on a stage with the plane of polarization in the west–east direction. When will the crystal reach extinction?

   The unit cell of a triclinic crystal has three axes of unequal length and no axes are perpendicular. The optic axes are rarely perpendicular to a crystal face, and are mostly unrelated to the crystallographic axes. In general, a triclinic crystal will exhibit an intermediate RI value, not primary RIs. Upon rotation, extinction may be seen (typically with the crystal at some angle to the plane of polarization, but exhibiting an intermediate refractive index).

CHAPTER 9

1. In the pharmaceutical industry, it is common for new drugs to be screened for polymorphic forms by melting a pure solid and then observing crystal forms. No solvents are used. Why?

No solvents are used because they become part of the crystal (as in a hydrate or a co-crystal). If the solvents are not used in making the drug form to be marketed, the solvent should not be used to study possible polymorphisms.

2. For each of the following solids that might be encountered in an unknown white powder, develop a simple precipitation (or other observable) reaction and describe how you would perform it including method (as per Table 9.2) and reagents. Justify your selections.

   There are many possible reasonable answers for this question. The critical issue is formation of a solid with a distinctive crystal form. As an example:

Solid	K_{sp}	Reagent	Reaction
$CaCO_3$	4.5×10^{-9}	HCl	Formation of CO_2 gas
$CaSO_4$	2.4×10^{-5}	Nitrate of barium, mercury, or lead	Solid formation

3. What are some of the practical limitations to microchemical analysis of unknowns such as white powders?

   The amount of sample available, homogeneity, relative concentrations of ions, metals, and constituents of concern, and the presence of interfering materials.

Appendix 2: Abbreviations

B or Bi: Birefringence

CCD: Charge coupled device; sensor in a camera that helps detect a light signal and convert it into a charge that is read as a digital signal

DOF: Depth of field

D-SLR: Digital single lens reflex camera

E: Extraordinary ray

EMS: Electromagnetic spectrum

EW: East–west direction convention for polarized light microscopy

F: Fast component or focal length

FOV: Field of view

K: Generic equilibrium constant

K_a: Acid dissociation constant

KI: Kohler Illumination

K_{sp}: Solubility product constant

M: Magnification

ML: Michel-Levy

n or RI: Refractive index

NA: Numerical aperture

NS: North–south direction convention for polarized light microscopy

O: Ordinary ray

pK_a: The $-\log(K_a)$; often used for calculation and comparison of acid strength

PLM: Polarizing light microscopy

Q: Calculated value of an equilibrium constant value at non-equilibrium conditions

RGB: Red-green-blue image; manner of saving a color image; each color is represented by a separate channel (image file)

RI or n: Refractive index

ROI: Region of interest; area of an image selected for application of a filter or transformation

S: Slow component

SLR: Single lens reflex camera; viewing occurs through the main lens; in most modern SLR cameras, the light of the object is also measured through the lens (TTL) to ensure correct exposure

TIFF: Tagged image file format; standard image format

USB: Universal serial bus; standard for connecting devices to a computer

Appendix 3: Glossary of Terms

Abbe: Ernst Abbe who developed the Abbe condenser and the Abbe theory of imaging.

Abbe condenser: Commonly constructed of two lenses; uncorrected for spherical and chromatic aberrations.

Aberration: Effect that prohibits a lens from focusing at a specific point.

Achromat: Objective corrected for chromatic aberrations (red and blue) and for spherical aberrations at a single wavelength.

Airy disk: Diffraction pattern of a point source formed by a perfect optical system with a circular aperture.

Analyzer: In polarizing light microscopy, a polarizing filter placed after the sample and before the eyepiece.

Anistotropic: Not the same; material that has more than one refractive index; light does not propagate with the same velocity in all directions.

Aplanatic–achromatic condenser: Condenser corrected for chromatic and spherical aberrations.

Apochromat: Highest quality objective with chromatic corrections at four wavelengths and spherically corrected at three wavelengths; generally has highest numerical aperture.

Astigmatism: Aberration that causes light in different planes to focus at different points, e.g., rays in the yz plane focus at a different point from rays in the xy plane.

Attenuation: Change in intensity or amplitude of light.

Back focal point: Focal point on image side of a lens.

Bertrand lens: Intermediate lens inserted into the tube of a microscope to image back focal plane of objective (and conjugate planes of illumination light path).

Biaxial crystal: Crystal with two optic axes and three refractive indices.

Birefringence: Numerical difference between two refractive indices in an anisotropic material such as calcite; may be positive or negative.

Black body: Object that absorbs all radiation incident upon it.

Bright field: Basic illumination system of a microscope; a light cone from the condenser is used to image the object.

Brightness: Absolute intensity of smallest part of an image.

Bulk sample: A sample as it arrives in a laboratory, from which representative samples are to be collected.

Chromatic aberration: Effect caused when white light passing through a lens has different points of focus depending on wavelength.

Co-crystal: Crystal containing more than one type of ion or compound (ionic or molecular).

Coma: Variation of magnification with aperture; off-axis peripheral streak of light.

Condenser: Optical component designed to provide a cone of light to illuminate a specimen.

Condenser annulus: Device that allows only a ring of light to pass through a condenser.

Coning and quartering: Systematic procedure for mixing samples to ensure homogeneity.

Conjugate planes: In Kohler Illumination, a set of conjugates planes that forms rays and illuminates an image; for each set, the specimen or filament can be imaged at all of the conjugate planes of the respective ray set.

Contrast: Variation of light intensity within an image.

Critical illumination: See Nelsonian illumination.

Crystal growth: Second stage of solid formation when ions add to a nucleation particle.

Dark field: Technique applied to improve contrast; only light reflected or scattered by a specimen is used to form an image.

Depth of field: The "thickness" of a sample in which all components are in focus.

Diffraction: Interaction of light with a limiting edge; bending of light at an edge; occurs when a wavefront passes through a small aperture, resulting in disruption of the wavefront.

Diffraction colors: After adjustment of the focus of a condenser (especially with high magnification objectives), a range of colors is seen as the working distance changes.

Diffraction grating: Series of transparent and opaque lines whose period is close to that of the wavelength of light.

Diopter: Reciprocal of focal length (in meters).

Direction of propagation: Direction in which a wave of light travels.

Direction of vibration: Plane in which a component of electromagnetic energy (here, the electrical component) vibrates.

Distortion: When the position of a specimen point is off its predicted position, the image is distorted; stated in terms of an item's relative deviation from its predicted position.

Double refraction: Splitting of light into ordinary (O) and extraordinary (E) components due to direction-dependent differences in refractive indices; calcite displays this property.

D wave: In phase contrast microscopy, wave that undergoes diffraction.

Electromagnetic energy: Energy that propagates through space and has electrical and magnetic components.

Ellipsoid of revolution: Ellipse rotated about an x or y axis to form a solid.

Extinction: In polarized light microscopy, point of rotation of a sample where no light reaches the eyepiece.

Extinction angle: In polarized light microscopy, angle of rotation of a birefringent sample where all light is blocked by the analyzer.

Extraordinary ray (E): In a uniaxial crystal, rays other than the main (or ordinary) are called extraordinary or epsilon (ε); in a birefringent material, the component whose velocity varies with direction.

Eyepoint: Height at which the eye should be above the eyepiece.

Fast component (F): In a birefringent material, component of the split beam that travels the fastest.

Field curvature: A positive lens forms a curved image; this effect can be corrected by special lens combinations.

Firewire: Type of serial bus standard; also known as IEEE 1394 interface.

Focal length: Measurement of a lens determined by its radius of curvature and refractive index; distance from the lens to the point where parallel rays, after entering the lens, converge after exit.

Front focal point: Focal point on object side of lens.

Illuminating rays: Set of rays used in Kohler Illumination to describe the filament at various points in the optical system.

Image forming rays: Set of rays used in Kohler Illumination to describe the image of a specimen at various points in the optical system.

Indicatrix: A diagram illustrating the combination of refractive indices in a crystal.

Interference: Interaction of separate waves of light; may be constructive or destructive.

Interference colors: Colors produced as a result of retardation in birefringent material.

Interpupillary distance: The distance between the pupils of eyes; must be set on the binocular head of a microscope.

Ionizable center: Ionizable group within a molecule such as carboxylic acid, phenol, or amine.

Isotropic: All the same; material that has only one refractive index, e.g., glass and sodium chloride; light within the material propagates with the same velocity in all directions.

Isotropic crystal: Crystal with only one refractive index.

JPEG format: Joint Photographic Experts Group's standard image file format (JPEG File Interchange Format [JFIF]).

Kohler Illumination: Illuminating technique to ensure even illumination of sample; ideal for photomicrography.

Lossy compression algorithm: A process to reduce size of (compress) an image file by reducing the amount of data retained in final image.

Magnification: Size of an image relative to its actual size.

Major axis: Longer axis of ellipse.

Michel-Levy chart: Chart showing interference colors and orders that relate birefringence to sample thickness.

Micrometer: Microscope slide with calibrated markings and known distances used to calibrate markings on an eyepiece.

Minor axis: Shorter axis of ellipse.

Monochromatic: Light of single wavelength.

Negative distortion: Barrel distortion.

Nelsonian illumination: Image appearing when a light source is focused in a sample plane; sources like filaments interfere with the image; rarely used.

Nosepiece: Device that holds a number of objectives and, upon rotation, introduces a new objective into the optical path.

Nucleation: Initial stages of crystal formation when a few ions combine in a crystal lattice.

Numerical aperture (NA): Cone of light gathered by lens; depends on size of lens and distance between object and lens.

Objective lens: Multiple lenses combined in the barrel of an objective.

Occlusion: Trapping of inert or foreign materials inside a crystal lattice.

Optic axis: Within a crystal, a direction of propagation in which fast and slow components of plane polarized light exhibit same velocity; in a birefringent sample, the O ray will follow the optic axis.

Optic axis angle: Acute angle between two optical axes in a biaxial crystal.

Optically negative crystal: $\varepsilon < \omega$ for a uniaxial crystal and $\beta - \alpha < \beta - \gamma$ for a biaxial crystal.

Optically positive crystal: $\varepsilon > \omega$ for a uniaxial crystal and $\beta - \alpha < \beta - \gamma$ for a biaxial crystal.

Ordinary ray (O): Ray that travels down the main axis in a uniaxial crystal in which the local environment of the ordinary ray is isotropic; also called omega (ω); in a birefringent material, component that follows normal rules of refraction.

Particle growth: Same as crystal growth.

Pentaprism: Five-sided prism used in a camera to view the image through the lens; light path is deviated by 90° and image remains upright.

Phase contrast: Technique utilizing phase shifts of objects to enhance their contrast.

Phase objects: Objects that do not produce inherent contrast; they simply cause a phase shift in light passing through them.

Photomicrography: Capturing an image through a microscope; microphotography is taking very small photographs.

Photon: Discrete packet of electromagnetic energy.

Pixel: Picture element; smallest detail in a digital image.

Plano-convex lens: Lens with one planar side and one convex side.

Polarizer: In polarizing light microscopy, a polarizing filter placed between light source and sample.

Polychromatic: Many colors; describes light sources that emit a range of wavelengths, e.g., a light bulb used as a source of illumination in a microscope.

Polymorphic: Many forms; compounds that have more than one stable crystal form.

Positive distortion: Pincushion distortion.

Prewitt operator: Filter designed to detect edges of an image.

Pseudopolymorph: Structure such as a hydrated crystal.

P wave: In phase contrast microscopy, a combination of S and D waves.

Raw format: Image acquired by the sensor of a camera, with little or no processing.

Rayleigh criterion: Assumes that two points can be resolved if separated by the diameter of the first dark ring in an Airy disk.

Real image: Image capable of being projected onto a screen.

Refractive index: Ratio of the velocity of radiation in a vacuum relative to the velocity of radiation in the material.

Representative sample: Subsample of a larger bulk sample whose composition accurately reflects and represents that of the bulk sample.

Resolution: Ability to establish separation of two very close objects.

Retardation: One component of a split beam in a birefringent material falls behind the other.

Semi-apochromat: Objective with better correction for chromatic aberrations than provided by an achromat; corrected for spherical aberrations at two wavelengths; also known as fluorite objectives.

Slow component (S): In a birefringent material, the component of a split beam that travels the slowest.

Snell's law: As a light moves from a medium with a relatively small refractive index into a medium of relatively high refractive index, it will be bent toward the normal in the medium with a higher index of refraction.

Solubility rules: Rules that determine whether a particular cation–anion combination will result in formation of an insoluble precipitate.

Spectrometer: Instrument that determines properties of a substance as a function of the wavelength of the incident light.

Specular reflection: Mirror-like reflection in which the angle of the reflected light is the same as the angle of the incident light relative to a line drawn perpendicular to the surface (the surface normal).

Spherical aberration: Rays that travel through the center of a lens and those traveling through the outer edges have different path lengths through the lens and therefore do not all focus at exactly the same point.

Substage condenser: Optical component designed to provide a cone of light to illuminate a specimen.

Supersaturated: Solution in which the concentrations exceed equilibrium concentrations.

S wave: In phase contrast microscopy, a wave that undergoes no diffraction.

Triaxial ellipsoid: Solid ellipse in which the sizes of the x, y, and z axes are different.

Trinocular head: Microscope head that allows for two eyepieces and a camera mount; some include a beamsplitter that separates the beam 50/50 into eyepieces and camera, or a mirror that directs the light to the eyepieces or the camera.

Uniaxial crystal: Crystal with one optic axis and two refractive indices.

Unit cell: Basic unit of a crystal lattice.

Virtual image: Image that can be observed through a microscope but cannot be projected onto a screen.

Appendix 4: Further Reading and Resources

The following are general references and resources for microscopy. Other specific references are cited in the text as appropriate.

AMERICAN SOCIETY FOR TESTING AND MATERIALS STANDARDS

ASTM, Standard guide for microcrystal testing in the forensic analysis of cocaine, 2003.
ASTM. Standard guide for microcrystal testing in the forensic analysis of methamphetamine and amphetamine, 2006.
ASTM, Standard guide for microcrystal testing in the forensic analysis of phencyclidine and its analogues, 2007.

USEFUL WEBSITES

U.S. Naval Research Laboratory, *Crystal Lattice Structures* [cited November 23 2008]. Available from http://cst-www.nrl.navy.mil/lattice/.
Nikon Corporation, *Michel-Levy Interference Chart* [cited November 23 2008]. Available from http://www.microscopyu.com/articles/polarized/michel-levy.html.

BOOKS

R. Weast, Ed., CRC *Handbook of chemistry and physics*, 64th ed., 1983.
A. Abramowitz, *Microscope basics and beyond,* Vol. 1, Olympus America, 2003.
P.W. Atkins, *Physical chemistry*, 2nd ed., Oxford University Press, 1983.
W. Burger and M.J. Burge, *Digital image processing: An algorithmic introduction using Java*, Springer, 2008.
E.M. Chamot and C.W. Mason, *Handbook of chemical microscopy*, 2nd ed. (2 vols.), John Wiley & Sons, 1948.
E.M. Chamot and W.M. Mason, *Handbook of chemical microscopy* (2 vols.), McCrone Research Institute, 1931.
E.S. Dana, *Textbook of mineralogy*, 4th ed., John Wiley & Sons, 1932.
R. de Levie, *Advanced Excel for scientific data analysis*, Oxford University Press, 2004.
J.G. Delly, *Photography through the microscope*, 9th ed., Eastman Kodak, 1988.
R.W. Ditchburn, *Light*, 2nd ed., Blackie & Son, 1963.
F.M. Ernsberger, *Polarized light in glass research*, PPG Industries, 1970.
R.E. Fischer, B. Tadic-Galeb, and P.R. Yoder, *Optical system design*, 2nd ed., SPIE, 2008.
C.C. Fulton, *Modern microcrystal tests for drugs*, Wiley Interscience, 1969.
R.C. Gonzalez and R.E. Woods, *Digital image processing*, 3rd ed., Prentice Hall, 2007.
D. Halliday, D and R. Resnick, *Fundamentals of physics*, 2nd ed., John Wiley & Sons, 1981.
N.H. Hartshorne and A. Stuart, *Crystals and the polarizing microscope*, 4th ed., American Elsevier, 1970.
D.S. Kliger, J.W. Lewis, and C.E. Randall, *Polarized light in optics and spectroscopy*, Academic Press, 1990.

W.C. McCrone, L.B. McCrone, and J.G. Delly, *Polarized light microscopy*, McCrone Research Institute, 1995.

D.B. Murphy, *Fundamentals of light microscopy and electronic imaging*, Wiley-Liss, 2001.

D. O'Shea, *Elements of modern optical design*, John Wiley & Sons, 1985.

A. Peck, *Beginning GIMP*, Apress, 2006.

N. Petracco and T. Kubic, *Color atlas and manual of microscopy for criminalists, chemists, and conservators*, CRC Press, 2003.

G. Reis, *Photoshop CS3 for forensics professionals*, Sybex, 2007.

J. Robertson, *Forensic examination of hair*, Taylor & Francis, 1999.

J.C. Russ, *Image processing handbook*, 5th ed., CRC Press, 2007.

E.J. Spitta, *Microscopy*, 3rd ed., John Murray, 1920.

R.E. Stoiber and S.A. Morse, *Microscopic identification of crystals*, Robert E. Krieger, 1981.

B.P. Wheeler and L.J. Wilson, *Practical forensic microscopy*, Wiley-Blackwell, 2008.

SELECTED JOURNAL ARTICLES

K.M. Andera, H.K. Evans, and C.M. Wojcik, Microchemical identification of γ-hydroxybutyrate (GHB), *Journal of Forensic Sciences* 45: 665–668, 2000.

S.C. Barthe, M.A. Grover, and R.W. Rousseau, Observation of polymorphic change through analysis of FBRM data, *Crystal Growth and Design* 8: 3316–3322, 2008.

H.G. Brittain, Foundations of chemical microscopy 1: Solid-state characterization of 5-nitrobarbituric acid (dilituric acid) and its complexes with Group IA and Group IIA cations, *Journal of Pharmaceutical and Biomedical Analysis* 15: 1143–1155, 1997.

H.G. Brittain, Foundations of chemical microscopy 2: Derivatives of primary phenylalkylamines with 5-nitrobarbituric acid, *Journal of Pharmaceutical and Biomedical Analysis* 19: 865–875, 1999.

H.G. Brittain, Polymorphism and solvatomorphism, *Journal of Pharmaceutical Sciences* 96: 705–728, 2007.

H.G. Brittain and M. Rehman, Foundations of chemical microscopy 3: Derivatives of some chiral phenylalkylamines and phenylalkylamino acids with 5-nitrobarbituric acid, *Chirality* 17: 89–98, 2005.

J. Cartier, C. Roux, and M. Grieve, A study to investigate the feasibility of using x-ray fluorescence microanalysis to improve discrimination between colorless synthetic fibers, 1997.

Y. Chang and H.C. Zeng, Manipulative synthesis of multipod frameworks for self-organization and self-amplification of Cu_2O microcrystals, *Crystal Growth and Design* 4: 273–278, 2004.

D.L. Chen, C.J. Gerdts, and R.F. Ismagilov, Using microfluidics to observe the effect of mixing on nucleation of protein crystals, *Journal of the American Chemical Society* 127: 9672–9673, 2005.

Y. Chen, Z.P. Guo, X.Y. Wang, and C.G. Qiu, Sample preparation, *Journal of Chromatography A* 1184: 191, 2008.

S.L. Childs et al., A metastable polymorph of metformin hydrochloride: Isolation and characterization using capillary crystallization and thermal microscopy techniques, *Crystal Growth and Design* 4: 441–449, 2004.

L.J. Chyall et al., Polymorph generation in capillary spaces: Preparation and structural analysis of a metastable polymorph of nabumetone, *Crystal Growth and Design* 2: 505–510, 2002.

C.L. Cooke and R.J. Davey, On the solubility of saccharinate salts and co-crystals, *Crystal Growth and Design* 8: 3483–3485, 2008.

K.H. Esbensen et al., Representative process sampling in practice: Variographic analysis and estimation of total sampling errors (TSE), *Chemometrics and Intelligent Laboratory Systems* 88: 41–59, 2007.

F.M. Garfield, Sampling in the analytical scheme, *Journal of the Association of Official Analytical Chemists* 72: 405–411, 1989.

M. Grieve and S. Deck, A new mounting medium for the forensic microscopy of textile fibers, *Science and Justice* 35: 109–112, 1995.

P.M. Gy, Introduction to theory of sampling 1: Heterogeneity of a population of uncorrelated units, *Trac-Trends in Analytical Chemistry* 14: 67–76, 1995.

J.O. Henck, J. Bernstein, A. Ellern, and R. Boese, Disappearing and reappearing polymorphs: The benzocaine/picric acid system, *Journal of the American Chemical Society* 123: 1834–1841, 2001.

J.L. Hilden et al., Capillary precipitation of a highly polymorphic organic compound, *Crystal Growth and Design* 3: 921–926, 2003.

J.B. Holm-Nielsen, C.K. Dahl, and K.H. Esbensen, Representative sampling for process analytical characterization of heterogeneous bioslurry systems: Reference study of sampling issues in PAT, *Chemometrics and Intelligent Laboratory Systems* 83: 114–126, 2006.

K. Kim, K.E. Plass, and A.J. Matzger, Kinetic and thermodynamic forms of a two-dimensional crystal, *Langmuir* 19: 7149–7152, 2003.

H.K. Lee, J.H. Kwon, S.H. Park, and C.W. Kim, Insulin microcrystals prepared by the seed zone method, *Journal of Crystal Growth* 293: 447–451, 2006.

L. Lewis and A.J. Sommer, Attenuated total internal reflection microspectroscopy of isolated particles: Alternative approach to current methods, *Applied Spectroscopy* 53: 375–380, 1999.

S.H. Long et al., Polymorphism of an organic system effected by the directionality of hydrogen-bonding chains, *Crystal Growth and Design* 8: 3137–3140, 2008.

K.R. Mitchell-Koch and A.J. Matzger, Evaluating computational predictions of the relative stabilities of polymorphic pharmaceuticals, *Journal of Pharmaceutical Sciences* 97: 2121–2129, 2008.

M. Navratil, G.A. Mabbott, and E.A. Arriaga, Chemical microscopy applied to biological systems, *Analytical Chemistry* 78: 4005–4019, 2006.

J.D. Ng, J.A. Gavira, and J.M. Garcia-Ruiz, Protein crystallization by capillary counterdiffusion for applied crystallographic structure determination, *Journal of Structural Biology* 142: 218–231, 2003.

G. Nichols and C.S. Frampton, Physicochemical characterization of the orthorhombic polymorph of paracetamol crystallized from solution, *Journal of Pharmaceutical Sciences* 87: 684–693, 1998.

M. Paakkunainen, S.P. Reinikainen, and P. Minkkinen, Estimation of the variance of sampling of process analytical and environmental emissions measurements, *Chemometrics and Intelligent Laboratory Systems* 88: 26–34, 2007.

L. Petersen and K.H. Esbensen, Representative process sampling for reliable data analysis: A tutorial, *Journal of Chemometrics* 19: 625–647, 2005.

L. Petersen, P. Minkkinen, and K.H. Esbensen, Representative sampling for reliable data analysis: Theory of sampling, *Chemometrics and Intelligent Laboratory Systems* 77: 261–277, 2005.

C.P. Price, A.L. Grzesiak, M. Lang, and A.J. Matzger, Polymorphism of nabumetone, *Crystal Growth and Design* 2: 501–503, 2002.

S.L. Price, Computational prediction of organic crystal structures and polymorphism, *International Reviews in Physical Chemistry* 27: 541–568, 2008.

N. Rodriguez-Hornedo and D. Murphy, Significance of controlling crystallization mechanisms and kinetics in pharmaceutical systems, *Journal of Pharmaceutical Sciences* 88: 651–660, 1999.

B. Rodriguez-Spong et al., General principles of pharmaceutical solid polymorphism: A supramolecular perspective, *Advanced Drug Delivery Reviews* 56: 241–274, 2004.

C. Sauter, K. Dhouib, and B. Lorber, From macrofluidics to microfluidics for the crystallization of biological macromolecules, *Crystal Growth and Design* 7: 2247–2250, 2007.

V. Schawaroch and S.C. Li, Testing mounting media to eliminate background noise in confocal microscope 3-D images of insect genitalia, *Scanning* 29: 177–184, 2007.

O. Simonsen, Crystal structures of hexaaquamagnesium diliturate dihydrate, $Mg(H_2O)_6$ (2^+) $C_4H_2N_3O$ (2^-) $\cdot$ $2H_2O$ and the isomorphous calcium salt, *Acta Chemica Scandinavica* 51: 861–864, 1997.

J.A. Springer and F.D. McClure, Statistical sampling approaches, *Journal of the Association of Official Analytical Chemists* 71: 246–250, 1988.

G.P. Stahly, Diversity in single- and multiple-component crystals: The search for and prevalence of polymorphs and co-crystals, *Crystal Growth and Design* 7: 1007–1026, 2007.

M.V. Warkentin, V. Stanislavskaia, K. Hammes, and R.E. Thorne, Cryocrystallography in capillaries: Critical glycerol concentrations and cooling rates, *Journal of Applied Crystallography* 41: 791–797, 2008.

K. Wiggins and P. Drummond, Identifying a suitable mounting medium for use in forensic fibre examination, *Science and Justice* 47: 2–8, 2007.

S. Wu et al., Microanalysis of individual silver halide microcrystals, *Scanning Microscopy* 7: 17–24, 1993.

M.K. Yadav et al., In situ data collection and structure refinement from microcapillary protein crystallization, *Journal of Applied Crystallography* 38: 900–905, 2005.

B. Zheng, C.J. Gerdts, and R.F. Ismagilov, Using nanoliter plugs in microfluidics to facilitate and understand protein crystallization, *Current Opinions in Structural Biology* 15: 548–555, 2005.

Index